Report of Investigations No. 114

FORMATION OF THE WINK SINK, A SALT DISSOLUTION AND COLLAPSE FEATURE, WINKLER COUNTY, TEXAS

Robert W. Baumgardner, Jr., Ann D. Hoadley, and Arthur G. Goldstein

Assisted by Gary Hummel, D'nese Young, Melissa A. Sandstrom, and Jennifer Forman

Work performed under U.S. Department of Energy Contract No. DE-AC97-80ET46615

Bureau of Economic Geology • W. L. Fisher, Director

The University of Texas at Austin • Austin, Texas 78712 • 1982

Cover: Ground-level photograph of the Wink Sink, June 3, 1980. A section of the wall has collapsed into the hole, splashing water up to 30 ft (9 m) high. Photograph by John Weaver; reprinted courtesy of the *Winkler County News.*

CONTENTS

ILLUSTRATIONS

Figures

Tables

ABSTRACT

The Wink Sink in Winkler County, Texas, formed on June 3, 1980. Within 24 hours it had expanded to a maximum width of 360 ft (110 m). On June 5, 1980, maximum depth of the sinkhole was 110 ft (34 m), and volume about 5.6 million ft^3 (158,600 m^3). Between June 3 and June 6, 1980, a large area bordering the south rim of the sink subsided about 10 ft (3 m) relative to the north side. Further subsidence of 1.456 ft (44.4 cm) occurred along the southern rim between July 19 and December 12, 1980.

A probable precursor of the sinkhole was a solution cavity that migrated upward by successive roof failures, thereby producing a collapse chimney filled with brecciated rock. Dissolution of salt in the Permian Salado Formation is inferred to have produced the solution cavity. Depth of the Salado ranges from 1,300 to 2,200 ft (396 to 670 m). Data on the size and initial depth of the solution cavity are unavailable.

The Salado Formation in the region contains several dissolution zones. Occurrence of dissolution in the middle of the Salado evaporite sequence may have resulted from ground-water flow along fractured anhydrite interbeds. Water may have come in contact with salt by downward movement from overlying aquifers or by upward movement from underlying aquifers under artesian pressure.

The Wink Sink lies directly above the Permian Capitan Reef, which contains water that is unsaturated with respect to sodium chloride. Hydraulic head of water from the reef is higher than the elevation of the Salado Formation but lower than the head in the Triassic Santa Rosa Formation, a near-surface fresh-water aquifer. Fracture or cavernous permeability occurs above, within, and below the Salado Formation, as indicated by the loss of circulation of drilling fluid in wells drilled near the sinkhole. Consequently, a brine-density-flow cycle may be operating: relatively fresh water moves upward under artesian pressure and dissolves salt; the denser brine moves downward under gravity flow in the same fracture system. Alternatively, downward flow of water from aquifers such as the Santa Rosa Formation or Quaternary sediments above the salt is also a possible explanation for dissolution. A plugged and abandoned well that was located within the circumference of the sinkhole may have provided a conduit for water movement.

Composition of water in the Wink Sink resembles that of water in nearby wells producing from the Quaternary alluvium and from the Triassic Santa Rosa Formation.

Hendrick well number 10-A was drilled in 1928 at a site now within the circumference of the sinkhole. The well, which initially produced about 80 percent water from the Permian Tansill Formation, was plugged with cement and abandoned in 1964. The well was not used for brine disposal. Over 12 million barrels of salt water produced from the Hendrick Field were disposed of by injection into the Permian Rustler Formation during 1961. Waterflood projects in the Hendrick Field began in 1963 and are still in progress.

Sinkholes similar to the Wink Sink occur in other areas of North America. Their morphology, associated strata, and mode of formation suggest that dissolution, brecciation, and surface subsidence commonly occur during their formation.

INTRODUCTION

The sudden formation of the Wink Sink on June 3, 1980, attracted widespread public attention through both national and local news media. In the days following the appearance of the sinkhole, there was much speculation regarding its development and the possibility that additional sinkholes might develop. This report, based on several months of intensive investigation (June through December, 1980), addresses some of the questions concerning the origin of the sinkhole.

On June 3, 1980, at about 9 a.m., a Harvard Construction Company crew was inspecting a Gulf Oil Company brine pipeline on the east side of section 41, block B-5, Public School Lands (PSL) survey, in Winkler County, Texas. The pipeline was leaking from a collar joining two sections of 24-inch-diameter (61-cm) pipe (Juan Garcia, Harvard Construction Co., personal communication, September 28, 1980). Splashing water attracted one crew member to a 20-ft-wide (6-m) hole in the ground about 100 ft (30 m) west of the pipeline. Large blocks of earth were collapsing into the hole, throwing water 30 ft (9 m) into the air. By noon, the diameter of the hole was about 100 ft (30 m). Rapid expansion of the oblong cavity ceased by the next morning, when its longest (east-west) dimension was about 360 ft (110 m) (fig. 1A).

The maximum depth of the hole on June 5, 1980, was 110 ft (34 m); average depth was estimated at 80 ft (24 m) on the basis of a line-and-plummet survey of the hole. Surface area of the sink was about 70,400 ft^2 (6,540 m^2), and its volume about 5.6 million ft^3 (158,600 m^3). Volume is more than twice that of the Cargill salt plant sinkhole near Hutchinson, Kansas, and almost three times the size of the Panning sinkhole in Barton County, Kansas (Walters, 1978).

Blocks up to 30 ft (9 m) long continued to fall into the hole at irregular intervals for several weeks (fig. 1B). Annular cracks that surround the hole extend up to 290 ft (88 m) from the southern edge. A large area bordering the south rim of the sinkhole subsided about 10 ft (3 m) relative to the north side within the first 3 days of movement (fig. 2A, B). This small grabenlike depression is bounded on the east and west by fissures up to 60 ft (18 m) long that are tangent to the hole (fig. 3A). Subsidence has been accompanied by faulting and lateral movement of the depressed block, as shown by abundant tension fractures (fig. 3B).

Development of the sinkhole had little effect on oil field operations. Workmen cut and rerouted a Shell Oil Company pipeline 6 inches (15 cm) in diameter that carried crude oil to storage tanks 1,000 ft (305 m) northeast of the sinkhole (fig. 2B). The brine pipeline originally being inspected by the Harvard Construction Company maintenance crew was broken by the expanding hole (fig. 2A). As a result, the oil wells producing the brine had to be shut down.

Workers for Petro-Lewis Company, an independent oil company, were attempting to circulate cement behind the liner in Hendrick well number 3-A, a producing oil well located about 500 ft (152 m) south of the sinkhole, when the sinkhole began to form. They plugged and abandoned the well on June 5, 1980, because of the proximity of tension fractures to the well (Mike Handren, Petro-Lewis Company, personal communication, August 14, 1980).

REGIONAL GEOLOGIC SETTING

Surficial Geology

The Wink Sink formed 2.5 mi (4.0 km) northeast of Wink, southwestern Winkler County, Texas (fig. 4). Winkler County is covered principally by Quaternary deposits that obscure bedrock formations. Along the Concho Bluff in the northeastern corner of the county, Cretaceous strata are exposed.

Surface drainage is poorly developed in Winkler County because surface sediments are highly permeable and rainfall is infrequent and usually localized. Mean annual precipitation is about 12 inches (30 cm), and the annual net lake surface evaporation rate exceeds 70 inches (178 cm) (Arbingast and others, 1976). Most precipitation collects in playas and other internally drained depressions. The Wink Sink is located on a line described by a group of these surface depressions that extends from west of Kermit to east of Wink (fig. 4).

Stratigraphy

The Delaware Basin of southeast New Mexico and West Texas is the western part of the Permian Basin province (fig. 5). It is separated from the Midland Basin to the east by the Central Basin Platform, a north-south-oriented structural high. The Delaware Basin is primarily filled with sedimentary rocks of Permian age. Uppermost Permian strata comprise the Ochoan Series (table 1). This series is composed of four formations. The lower two formations, the Castile and Salado, contain most of the evaporite deposits in the Delaware Basin. The upper two

Figure 1. Aerial and ground-level oblique photographs of the Wink Sink. (A) Oblique aerial photograph, June 5, 1980; north is to upper right. Depth to water surface is about 33 ft (10 m). Annular tension cracks surround the hole. A tangential crack (arrow) on the southeast side of the hole marks the eastern boundary of a zone of continued subsidence. All photographs, except where noted, are by Robert W. Baumgardner, Jr. (B) Slab failure with block collapsing into the sinkhole, June 3, 1980. Photograph by John Weaver, reprinted courtesy of *Winkler County News.*

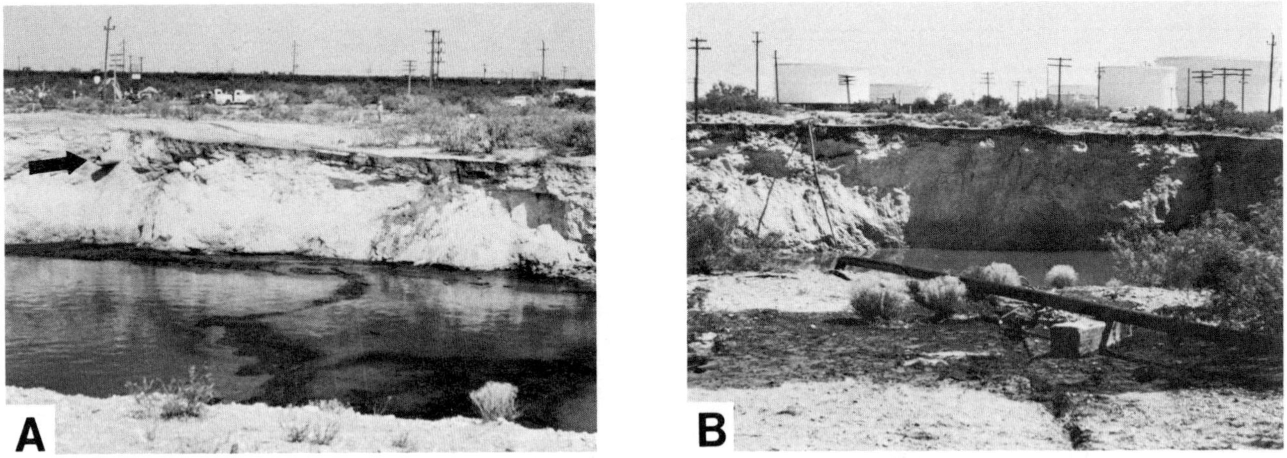

Figure 2. Local setting of the Wink Sink. (A) South side of the Wink Sink, June 5, 1980; view is to southeast from the northwest side of the hole. Area to the right (west) of brine pipeline (arrow) has subsided about 10 ft (3 m), producing a noticeable sag. (B) Wink sinkhole, June 5, 1980; view is to east from the southwest side of the hole. Pipeline in foreground was not in use when sinkhole formed. Oil pipeline on far side of sink was in use and had to be rerouted after being broken. Oil storage tanks in background are operated by Shell Oil Company.

Figure 3. Tension fractures near the Wink Sink. (A) Tangential surface fracture on east side of the subsided area, July 10, 1980; view is to north. Vertical displacement next to fieldbook (upper center of photograph) is about 18 inches (46 cm). This is the same crack marked by the arrow in figure 1A. (B) East-west tension fracture on south side of Wink Sink, November 18, 1980; view is to east. Maximum width of crack (24 inches [61 cm]) is a result of slumping. Block on left has moved only 6 inches (15 cm) to the left (north), relative to block on the right, as shown by width of crack adjacent to fieldbook.

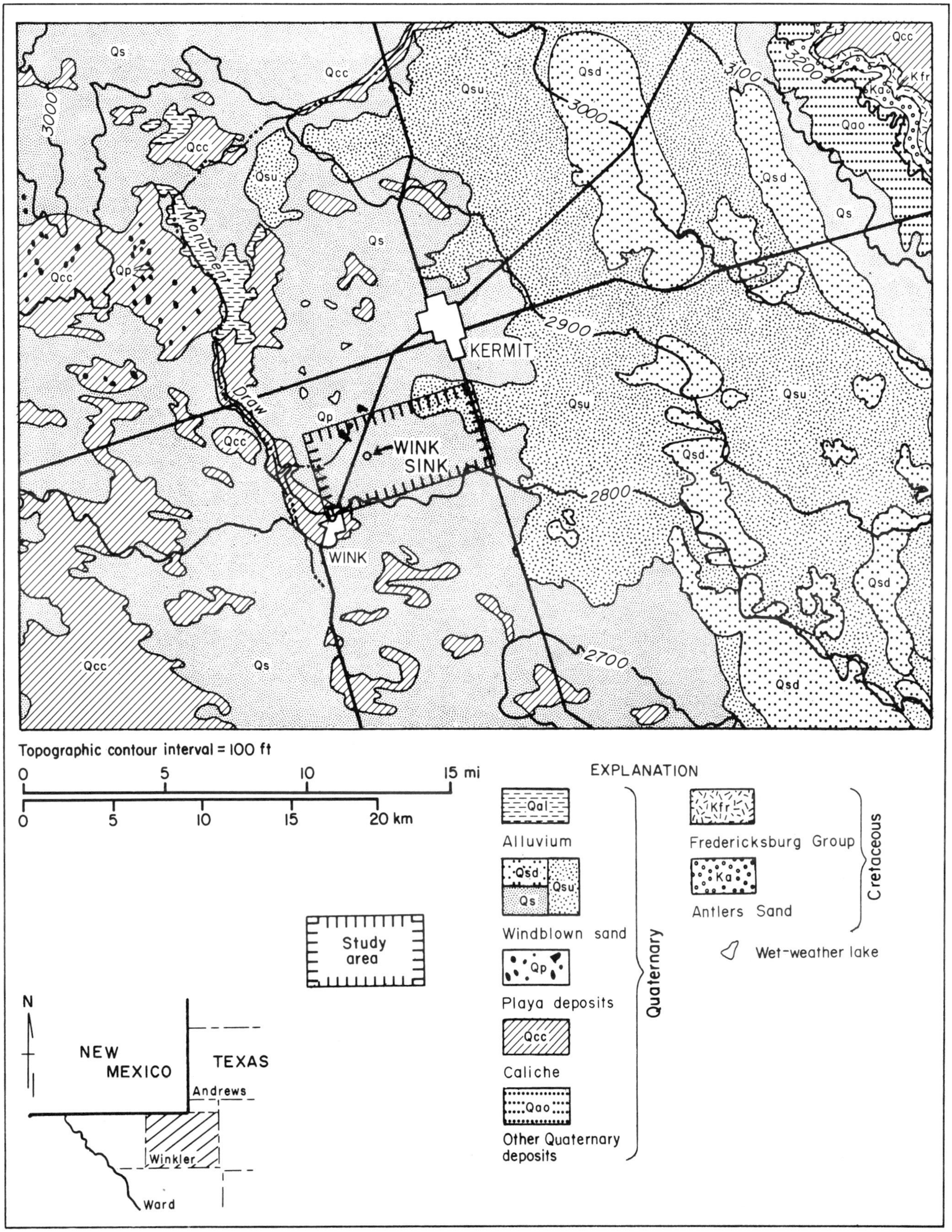

Figure 4. Generalized geologic map of southern Winkler County, Texas. The Wink Sink formed between Kermit and Wink in an area covered by windblown sand. The sinkhole lies along a line coincident with a group of northwest-southeast-trending wet-weather lakes. Study area refers to figures 8 through 10. Adapted from Barnes (1976).

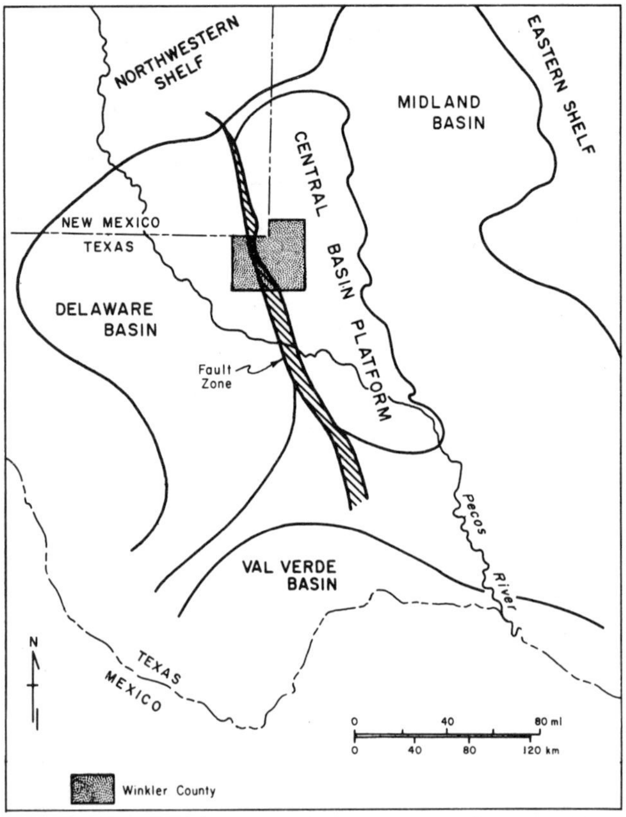

Figure 5. Structural setting of West Texas and eastern New Mexico. Regional tectonic element. Note fault zone coincident with western margin of the Central Basin Platform. Adapted from Hills (1970) and Keller and others (1980).

Table 1. Stratigraphic section examined in this report.

System	Series	Delaware Basin	Central Basin Platform
Quaternary		Alluvium	Alluvium
Triassic		Santa Rosa Tecovas	Santa Rosa Tecovas
Permian	Ochoa	Dewey Lake Rustler Salado Castile	Dewey Lake Rustler Salado
Permian	Guadalupe	Delaware Mountain Group: Bell Canyon Cherry Canyon Brushy Canyon	Capitan Reef — Artesia Group: Tansill Yates Seven Rivers Queen Grayburg San Andres Glorieta

Source: modified from West Texas Geological Society (1966).

formations, the Rustler and the Dewey Lake, are composed mainly of red beds, some gypsum and anhydrite, and minor amounts of salt (Johnson and Gonzales, 1978).

The Castile and the Salado Formations were deposited on uniform stable surfaces, as evidenced by the lateral continuity of individual laminae that can be traced over distances of several miles (Anderson and Kirkland, 1966; Hills, 1968; Anderson and others, 1972). The Castile Formation consists of anhydrite and halite, and may be as thick as 1,650 ft (503 m). It was deposited entirely within the Delaware Basin, bounded by the Capitan Reef, which essentially surrounds the basin (Anderson and Kirkland, 1980). The Castile contains more anhydrite than the overlying Salado Formation, which contains potash deposits (Johnson and Gonzales, 1978).

The Salado Formation was deposited over a larger area than was the underlying Castile, extending far beyond the Capitan Reef that defines the margin of the Delaware Basin. On the north and east sides of the basin (fig. 6), the Salado overlies the Capitan strata, but salt has been removed from the Salado Formation by dissolution in most of the area west of the Pecos River (Maley and Huffington, 1953). Where dissolution has not occurred, the Salado consists primarily of halite with some anhydrite interlayers. Depth to salt increases from approximately 165 ft (50 m) on the west side of the Delaware Basin to 2,540 ft (774 m) farther east. Maximum formation thickness reaches 1,950 ft (595 m) in the Delaware Basin; salt thickness reaches 1,650 ft (503 m) locally (Johnson and Gonzales, 1978). In the study area a maximum of 945 ft (288 m) of salt has been observed (table 2).

The Rustler Formation also contains salt. Individual salt beds are 6 to 33 ft (2 to 10 m) thick and constitute about 40 percent of the formation where dissolution has not occurred (Johnson and Gonzales, 1978).

The uppermost Permian formation, the Dewey Lake, contains no salt. It is composed primarily of red siltstone and some gypsum, anhydrite, and red shale (White, 1971) (figs. 6 and 7).

The Tecovas Formation of Triassic age unconformably overlies the Permian Dewey Lake Formation. In this report, the base of the Tecovas is defined as the contact between claystone of the Tecovas and siltstone of the Dewey Lake (appendix A).

Above the Tecovas Formation lies the Triassic Santa Rosa Sandstone, composed of medium- to coarse-grained sandstone (White, 1971). The Santa Rosa-Tecovas contact occurs between clean Santa Rosa sands and underlying fine-grained claystone of the Tecovas (appendix A). The contact of the Santa Rosa and overlying undifferentiated Triassic or Cenozoic strata is defined here as the first claystone or siltstone, or porosity break. Position of the upper and lower Santa Rosa contacts is approximate, as indicated by the dashed lines on figure 7. In the area covered by this study Triassic strata overlying the Santa Rosa Sandstone are not readily separable from Cenozoic sediments. Consequently, post-Santa Rosa deposits are not differentiated on cross sections (figs. 6 and 7). Where post-Santa Rosa Triassic sediments have been recognized, they are conformable with the underlying Santa Rosa Sandstone (White, 1971).

Structural Geology

The structural setting of the region is illustrated by a structure-contour map on the base of the lowermost dolomite within the Tansill Formation (fig. 8), which is predominantly dolomitic near the Capitan Reef (White, 1971). The Tansill Formation probably does not extend basinward beyond the top of the Capitan Reef (fig. 6). However, the dolomite used for constructing figure 8 can be traced throughout the study area (appendix A).

The Wink sinkhole is located above a closed structural high (fig. 8). Basal Tansill strata dip 25 ft/mi (4.7 m/km) to the east of the sinkhole and 500 ft/mi (95 m/km) to the west. The steeper westward dip probably reflects the steeply sloping face of the underlying Capitan Reef and is probably not related to salt dissolution.

Structural configuration on top of the Rustler Formation resembles that of the Tansill except that no structural high exists beneath the Wink Sink; however, a broad, closed high occurs 2 mi (3.2 km) to the east (fig. 9). East of the high, the formation dips eastward at about 75 ft/mi (14 m/km), about three times greater than the eastward dip at the base of the Tansill. West of the high, the Rustler dips 200 ft/mi (38 m/km) to the position of the Wink Sink and then increases dip to 500 ft/mi (95 m/km).

The westward dip exhibited by the Rustler Formation is greater than its eastward dip as a result of salt dissolution in the underlying Salado Formation. As the salt was removed, overlying formations collapsed to fill available space. Because the thickness of the Rustler does not increase into the dissolution trough (fig. 6), the trough was formed after Rustler deposition.

This hypothesis for the structure of the Rustler Formation can be substantiated by an isopach map of the Salado and Tansill Formations (fig. 10). The isopach map documents a decrease of more than 800 ft (244 m) in thickness from the thickest point shown on the map to the thinnest point in the dissolution trough on the west. These variations in thickness result from salt dissolution because neither the Tansill nor the anhydrite beds in the Salado vary more than 50 ft (15 m) in thickness (fig. 7).

Furthermore, the remarkable congruence between the isopach configuration of the Salado and Tansill Formations (fig. 10) and the structural configuration on top of the Rustler Formation (fig. 9), including isolated highs and reentrants, suggests that dissolution in the Salado strongly controls the structure of the Rustler. The Rustler appears to be draped over the underlying formation.

If solution cavities had developed in the Salado and migrated upward through the Rustler, closed depressions should occur at the top of the Rustler. Features of this kind are not apparent in figure 9, but the distance between most data points is too great to detect a feature less than 360 ft (110 m) in diameter, the approximate size of the Wink Sink.

Table 2. Thickness of Salado Formation from well logs used for cross sections B-B′ and D-D′.

		THICKNESS OF SALADO FORMATION		
	Well no.	Total (ft)	Salt (ft)	Anhydrite (ft)
CROSS SECTION B-B′	140	407	42	365
	139	666	306	360
	146	810	490	320
	147	780	480	300
	124	820	595	225
	87	950	700	250
	81	1,010	750	260
	77	1,140	880	260
	67	1,220	945	275
	66	1,240	925	315
CROSS SECTION D-D′	10	810	600	210
	109	625	415	210
	113	800	580	220
	163	830	600	230
	158	900	685	215
	155	860	635	225
	153	970	720	250
	269	970	710	260

SALT DISSOLUTION IN THE DELAWARE BASIN

History of Salt Dissolution

The chronology and geographic distribution of salt dissolution in the Delaware Basin were and are controlled by local hydrologic conditions and the geology of the basin. The timing and style of dissolution differ for the western and eastern parts of the basin.

In the western part of the basin, Salado salt deposits were dissolved when the Delaware Basin was tilted

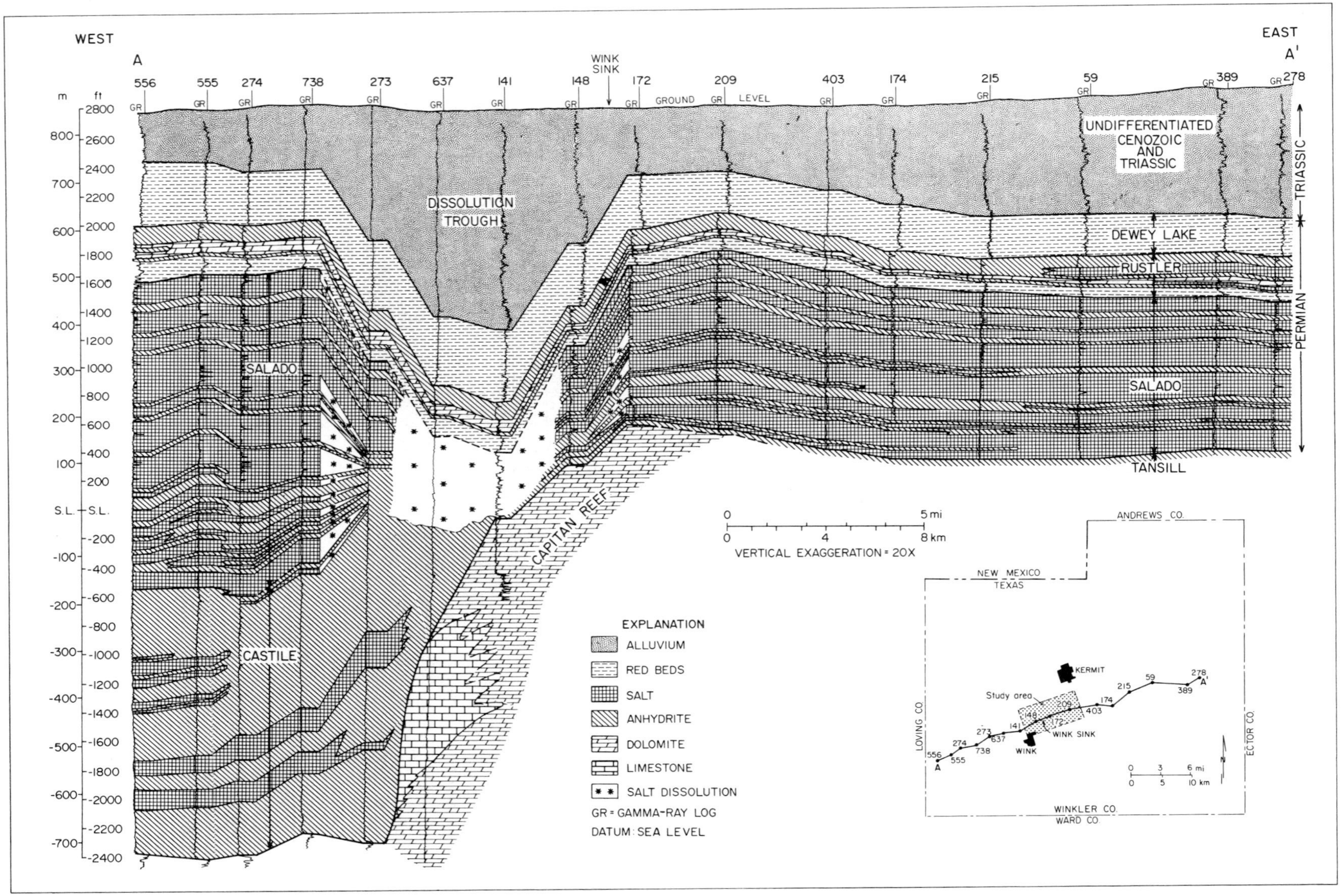

Figure 6. Regional east-west cross section A-A′ on the eastern edge of the Delaware Basin. Western flank of Central Basin Platform is marked by the Capitan Reef. Dissolution trough above Capitan Reef is result of salt dissolution in Salado and Castile Formations. Well numbers on all cross sections refer to appendix B. Study area on inset map (stippled) refers to figures 8 through 10.

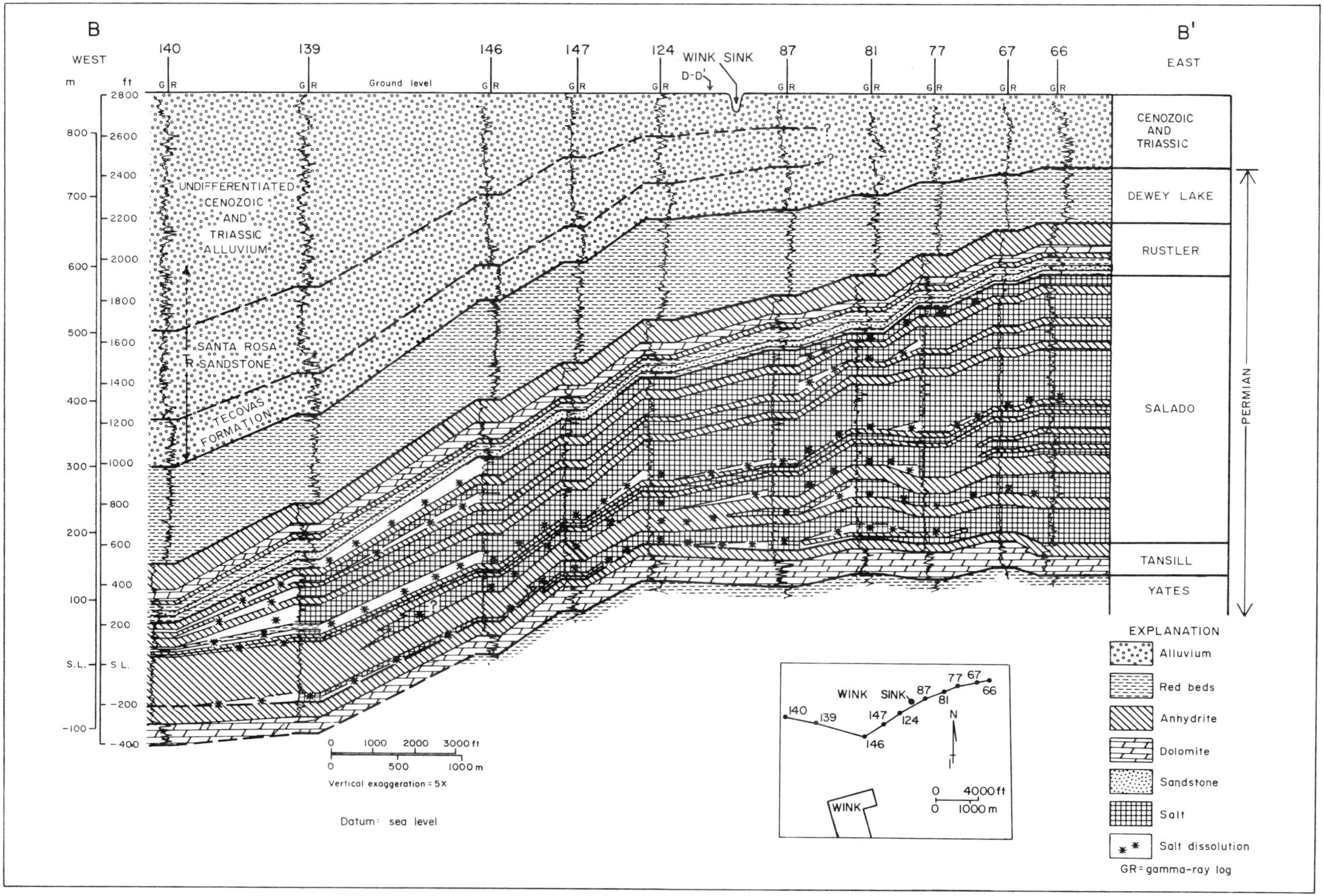

Figure 7. Local east-west cross section B-B′ showing salt dissolution zones in Salado Formation. Westward decrease in elevation of Permian and Triassic strata above the Salado is caused by dissolution and collapse in the salt-bearing section. Upper and lower contacts of Triassic Santa Rosa Sandstone are approximate. Inset shows location of cross section; line of section also shown on figures 8 through 10. See figures 15 and 16 for more detail.

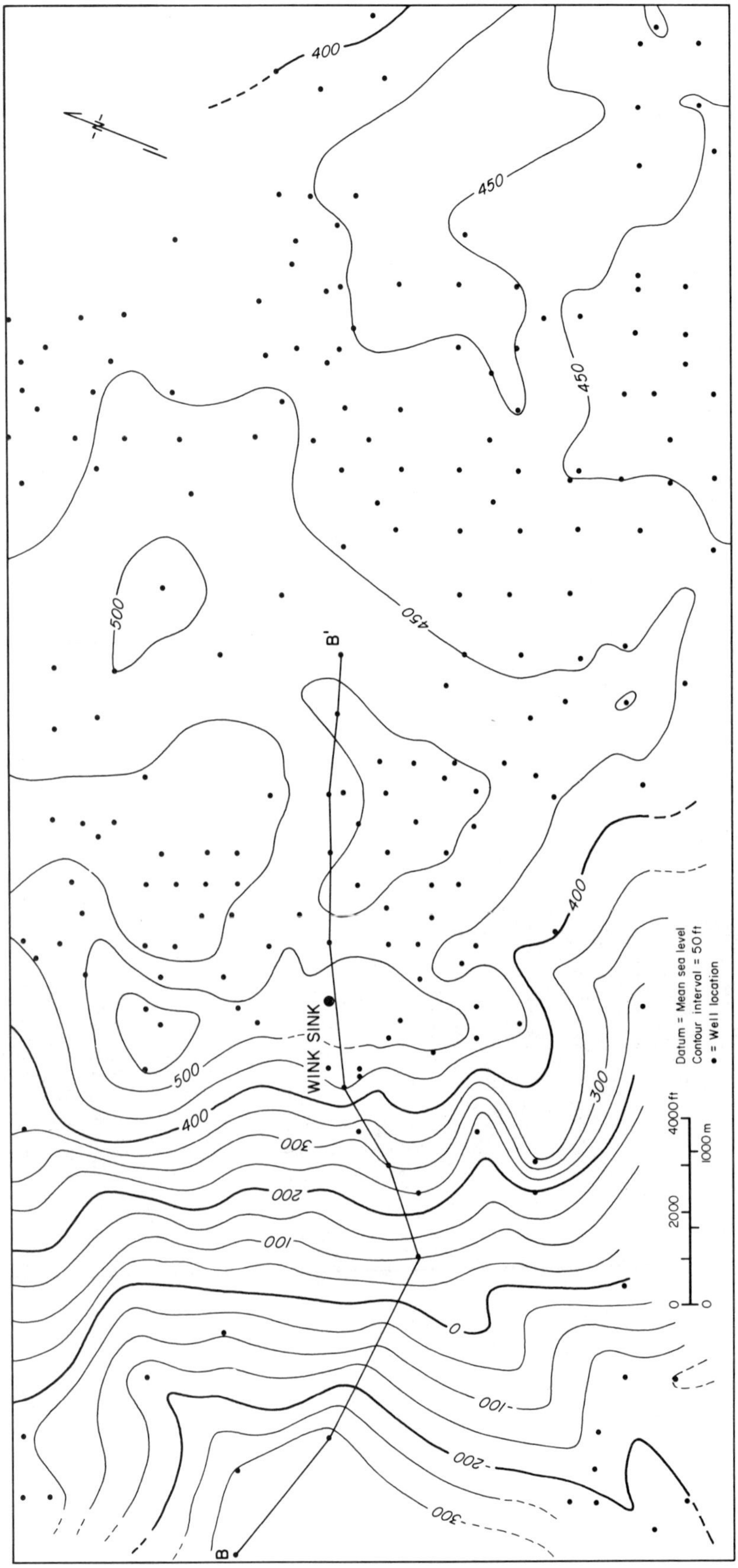

eastward between the end of Salado deposition and the beginning of Rustler deposition (Adams, 1944), and again in the late Tertiary (Maley and Huffington, 1953; Bodenlos, 1978). Structural tilting elevated strata on the western side of the basin, and as a result, salt was dissolved by ground and surface water coming into contact with either shallow or exposed salt beds (Mercer and Hiss, 1978).

In the eastern Delaware Basin, Salado salt dissolution and related subsidence began during the Permian (fig. 6) and probably continues today. The Permian Dewey Lake Formation is about 140 ft (43 m) thicker in the dissolution trough, indicating that its deposition coincided with or postdated a period of subsidence. Triassic and Cenozoic sediments are up to 1,100 ft (335 m) thicker in the trough (fig. 6), illustrating that dissolution and subsidence were active during the Triassic and Cenozoic.

Source of the waters that dissolved the Salado salt has been a subject of several geologic studies of the area. Maley and Huffington (1953) mapped the dissolution trough (fig. 6), and they ascribed the anomalous occurrence of 1,500 ft (457 m) of Triassic and Cenozoic alluvial deposits to dissolution of salt above the Capitan Reef aquifer. A correlation between salt dissolution and the Capitan Reef was earlier recognized by Adams (1944), who suggested that faults in the Rustler, caused by dissolution, subsidence, or warping of the underlying Capitan Reef, had facilitated movement of ground water down through the Rustler and into contact with Salado salts. Hills (1970) similarly concluded that dissolution may have been caused by ground water moving along joints and faults opened by

Figure 8. Structure-contour map on base of Tansill Formation. Wink Sink is located above a structural high that trends northwest to southeast. Base of Tansill dips westward at 500 ft/mi (95 m/km) and eastward at 25 ft/mi (4.7 m/km). For location of map, see figure 4 and figure 6 (inset). Line B-B' refers to cross section, figure 7.

movement along a north-south-trending fault zone on the west side of the Central Basin Platform (fig. 5).

Mercer and Hiss (1978), on the other hand, concluded that the Capitan Reef and shelf-aquifer systems were the source of the waters that dissolved Salado salt and formed the collapse features on the northeast side of the Delaware Basin (fig. 11). This hypothesis requires, of course, that the source of water be below the Salado salt. This mechanism was proposed by Parker (1967) to explain salt dissolution in the Williston Basin in North Dakota and in the Powder River Basin in Wyoming; more recently this hypothesis has been used by Anderson and Kirkland (1980) to explain dissolution in the Delaware Basin.

Mechanisms of Salt Dissolution

Brine-Density Flow

A brine-density-flow mechanism to explain dissolution by upward movement of unsaturated ground water in the Delaware Basin was recently described by Anderson and Kirkland (1980) (fig. 11). They report that the mechanism is a cycle with two components: (1) an underlying artesian source of relatively fresh water and (2) a permeable fracture zone between the underlying water source and salt strata that normally are isolated from shallow ground water.

Artesian pressure in the Capitan Reef aquifer is at least partly maintained by recharge from the Delaware Mountain Group (fig. 11). Water in the Delaware Mountain Group moves across the Delaware Basin from west to east (McNeal, 1965). Salinity of the water in

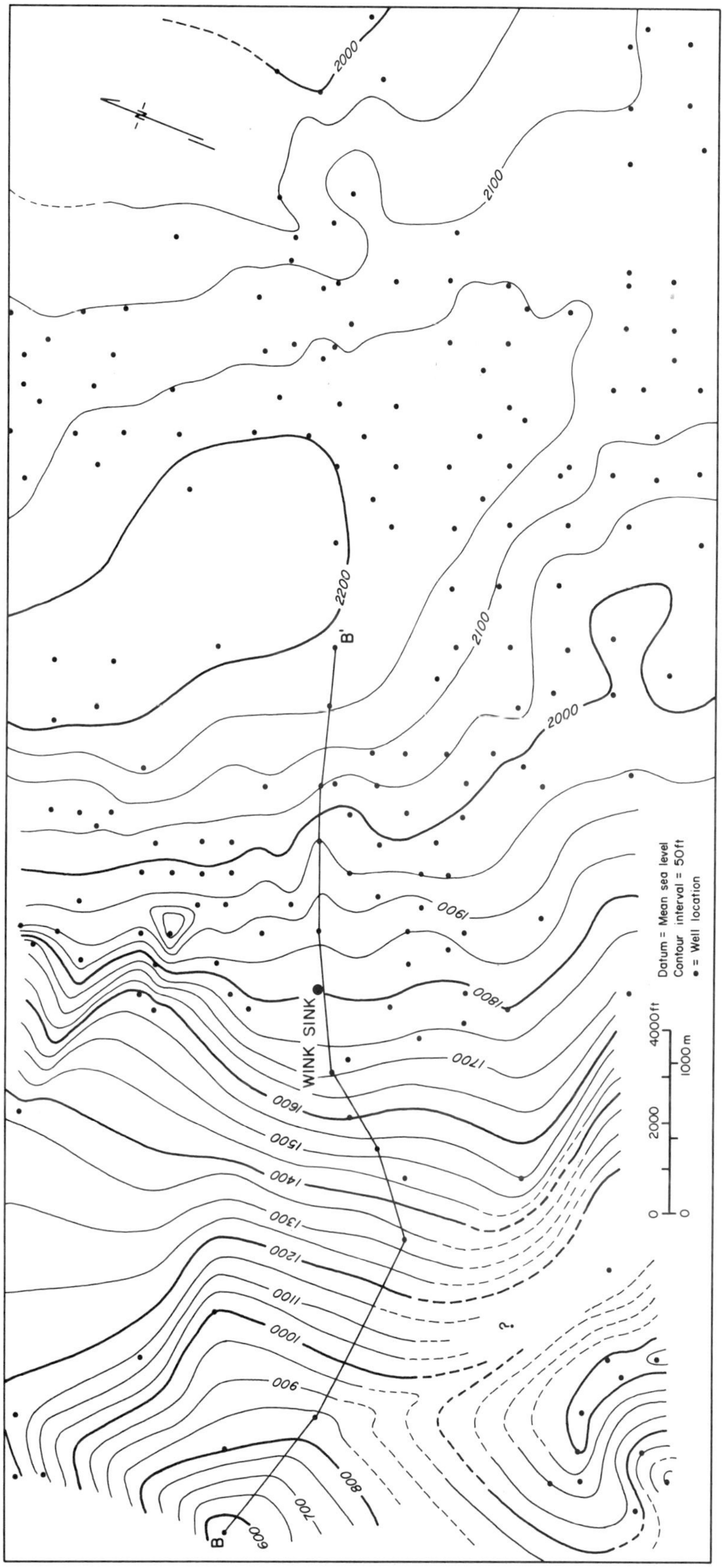

Figure 9. Structure-contour map on top of Rustler Formation. Top of the Rustler dips westward at 200 to 500 ft/mi (38 to 95 m/km) and eastward at 75 ft/mi (14 m/km) from northwest-southeast-trending structural high 2 mi (3.2 km) east of Wink Sink. For location of map, see figure 4 and figure 6 (inset). Line B-B' refers to cross section, figure 7.

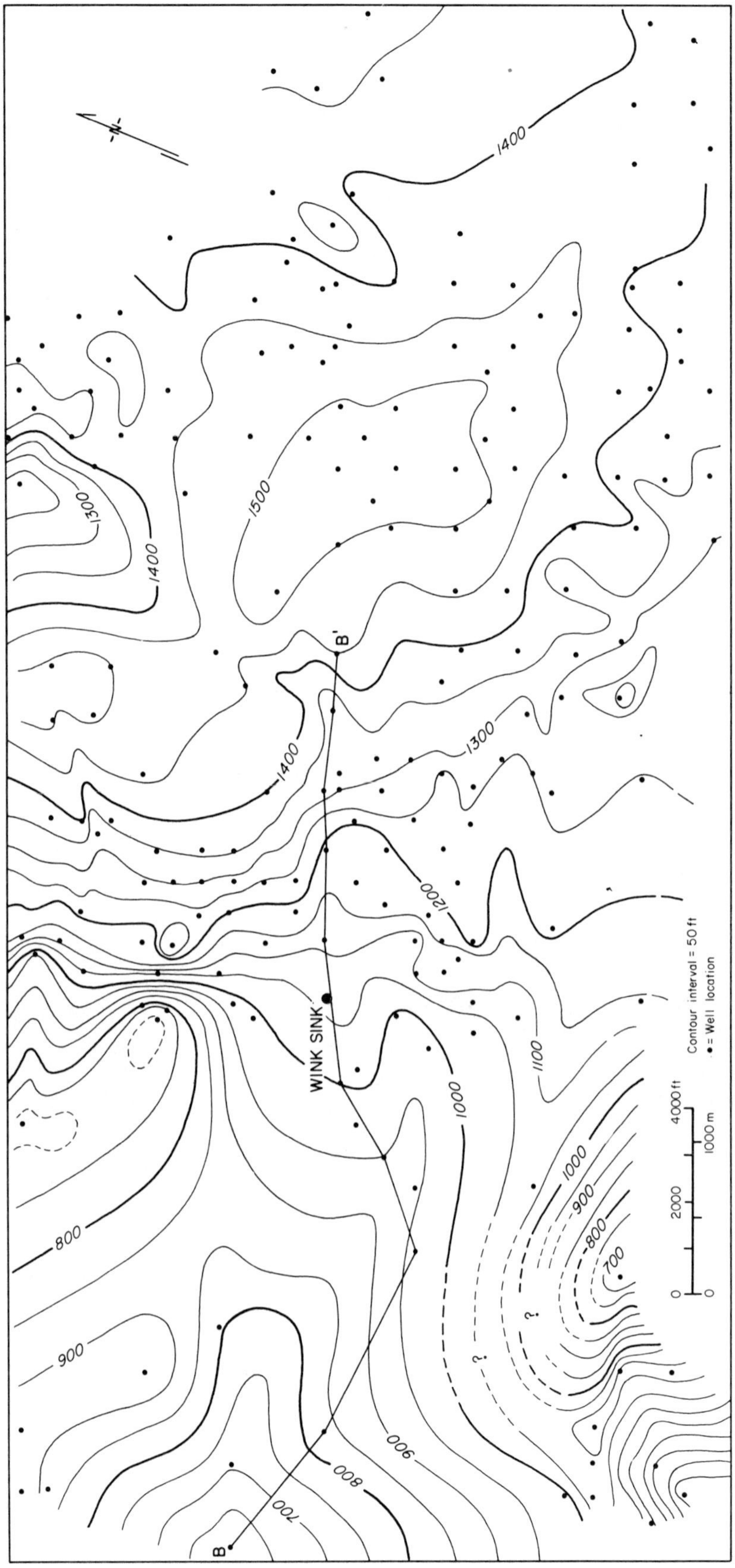

the Brushy Canyon, Cherry Canyon, and Bell Canyon Formations increases from less than 5,000 mg/L on the west to more than 200,000 mg/L on the east (Hiss, 1975).

Where the brine-density-flow cycle operates, relatively fresh water is forced upward under artesian conditions into contact with salt strata, where the salt is dissolved. Brine produced by dissolution is more dense than the fresher water; hence, the brine moves downward under gravity flow and forces fresher water to move upward to replace the brine, thereby perpetuating the cycle. Both downward and upward flow may occur simultaneously in the fracture system because of differences in fluid density.

Evidence of this phenomenon has been documented in southeastern Eddy County, New Mexico (Anderson and others, 1978). Dissolution zones were found in the upper part of the Castile Formation and lower part of the Salado Formation, but overlying salt beds in the upper and middle Salado were not dissolved.

A similar pattern of dissolution exists in western Winkler County, Texas. Salt dissolution zones occur at several levels within the Salado Formation and appear to be associated with anhydrite interbeds (fig. 7). The mechanism for dissolution in the middle of an evaporite sequence has not been fully explained, but Anderson and his co-workers (1978) suggest that permeable beds within the evaporite sequence could allow ground water to migrate into contact with and to dissolve salt beds. Anhydrite beds shown in figure 7 may be permeable pathways resulting from fracturing or partings between thin dolomite beds within the anhydrite. Fractures in the Salado Formation may have been caused by warping of these younger strata over the underlying Capitan Reef (Adams, 1944), by deeper solution and collapse, or by minor

Figure 10. Isopach map of the Salado and Tansill Formations. Variations in thickness are largely due to salt dissolution. For location of map, see figure 4 and figure 6 (inset). Line B-B' refers to cross section, figure 7.

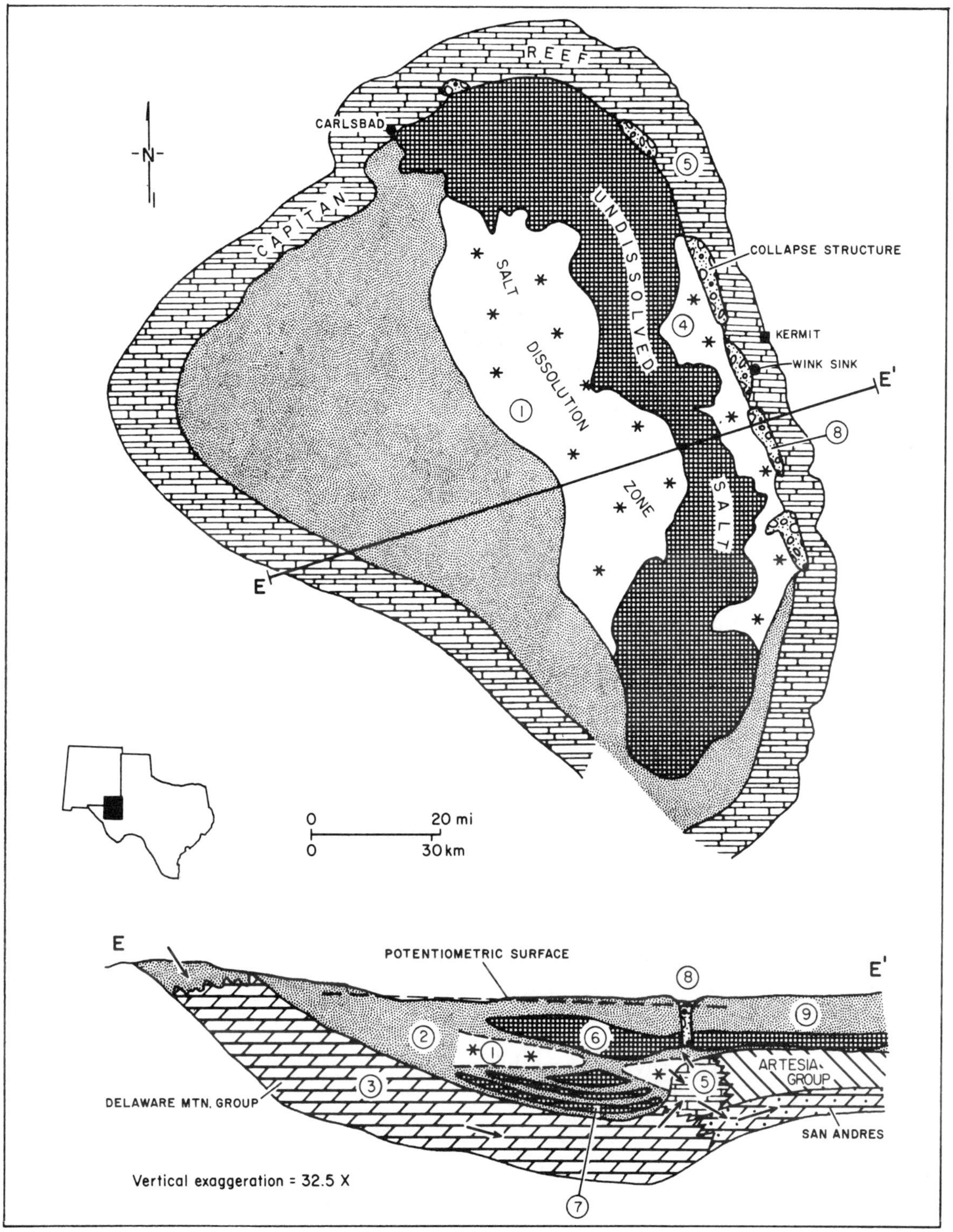

Figure 11. Schematic view of aquifer systems and salt dissolution, Delaware Basin. Path of water movement shown by arrows. Western dissolution zone (1) is result of tilting of Delaware Basin and subsequent dissolution of salt by downward-percolating ground water moving through formations (2) above the Delaware Mountain Group (3). Brine-density-flow model requires that eastern dissolution zone (4) was produced by water from Capitan Reef aquifer (5) that moved upward and laterally into Salado (6) and Castile Formations (7) under artesian pressure. Position of potentiometric surface is generalized. Salt dissolution leads to formation of collapse structures (8) that follow trend of inner margin of Capitan Reef. Location of Wink Sink is approximate. Post-Salado formations (9) east of the Capitan Reef probably affect dissolution only locally. Adapted from Anderson and Kirkland (1980).

faulting. Keller (1980) reports that microseismic events in the southwestern Winkler County and northern Ward County area are common.

Downward Ground-Water Flow

If permeable zones above and below the salt are interconnected, then the brine-density-flow mechanism is less feasible. Salt dissolution by the brine-density cycle functions only if salt strata are isolated from shallower aquifers that have hydrostatic heads higher than the underlying artesian aquifer (G. Fogg, personal communication, 1981). Results of drill-stem tests from wells near the Wink Sink show that in 1975 the hydraulic head in the Santa Rosa Formation was higher than that in the Tansill, Yates, or Capitan (fig. 12; table 3). If the Santa Rosa were connected with the aquifers underneath the Salado Formation via permeable zones, downward flow into the deeper aquifers would result; thus, the effect would be the reverse of the brine-density-flow model.

Furthermore, Garza and Wesselman (1959) report that in Winkler County, wells drilled into the Rustler Formation for waterflood projects yielded artesian water. Three wells completed in the Rustler north and northwest of Kermit between 1954 and 1957 had static water levels that were higher than those in the Yates and Tansill Formations in 1975 (Garza and Wesselman, 1959, table 7, pls. 1 and 3). No hydrologic data are available for the period between 1954 and 1957 from wells completed in the Rustler Formation near the location of the Wink Sink.

Two geologic factors, however, may impede downward movement of water from these two formations. First, the Dewey Lake Formation, which lies between the Santa Rosa and the Salado Formations, is a relatively impermeable red-bed sequence that acts as a barrier to water movement except where it is fractured, or perhaps, penetrated by wells. Second, water yields and permeability of the Rustler are highly variable, owing to the sporadic occurrence of cavernous porosity. Therefore, downward flow into the Salado Formation from the Santa Rosa and Rustler Formations may only occur locally. At the same time, brine-density flow may be occurring in isolated lower parts of the Salado Formation that are not hydraulically connected with the overlying aquifers.

Dissolution Phenomena in the Delaware Basin

Collapse Features

The proximity of several dissolution and collapse features to the Capitan Reef trend suggests that the reef facilitates their development (fig. 13). The Clayton Basin in eastern Eddy County, New Mexico, has subsided more than 100 ft (30 m) since the middle Pleistocene (Bachman, 1976). Nash Draw, an unfilled solution trough, is actively developing (Adams, 1944). According to Bachman (1976), parts of the draw have subsided as much as 180 ft (55 m) since middle Pleistocene.

Perhaps the best known solution features in the Delaware Basin are the San Simon Swale and the San Simon Sink in Lea County, New Mexico (fig. 13). The swale is a 100-mi^2 (260-km^2) elongate depression, trending northwest-southeast on the northeastern edge of the basin. The swale overlies and is parallel to the inner margin of the Capitan Reef. According to Nicholson and Clebsch (1961, p. 14), the swale "probably originated as the result of a deep-seated collapse."

The lowest part of the swale is the San Simon Sink. It covers an area of 0.5 mi^2 (1.3 km^2). The sink is about 130 ft (40 m) deep and is filled with at least 400 ft (120 m) of alluvium deposited above the surface of Triassic red beds. The most recent subsidence at the sink occurred about 50 years ago (Nicholson and Clebsch, 1961).

Several workers studied the relation between evaporite dissolution and ephemeral streams. Morgan (1942) believes that evaporite dissolution influenced the effectiveness and location of ephemeral stream channels in the Pecos River basin. Maley and Huffington (1953) suggest the opposite, that ephemeral surface drainage affected subsurface evaporites. They conclude that dissolution was probably enhanced along stream courses by percolation of fresh water into underlying sediments.

Monument Draw (fig. 13), which extends from central Lea County, New Mexico, to the Pecos River in Ward County, Texas, has been cited as an example of surface drainage that has been affected by evaporite dissolution (Nicholson and Clebsch, 1961). In southern Lea County, the north-south trend of the draw defines an acute angle to regional northwest-southeast dip, suggesting that its orientation results from stream capture by surface lowering along a trend parallel to the subsurface Capitan Reef. In northern Winkler County, the draw veers 90 degrees to the southwest, passing above the Capitan Reef, then abruptly changes direction again and passes 3.1 mi (5 km) west of the Wink Sink parallel with the trend of the inner margin of Capitan Reef (figs. 4 and 13). These angular bends in the draw may be controlled by subsurface faults or linear dissolution zones.

Pecos River Salt Load

The Pecos River is a discharge zone for saline springs in the Delaware Basin. The largest concentration of springs is at Malaga Bend in southeastern Eddy County, New Mexico (fig. 13). Historically, brine

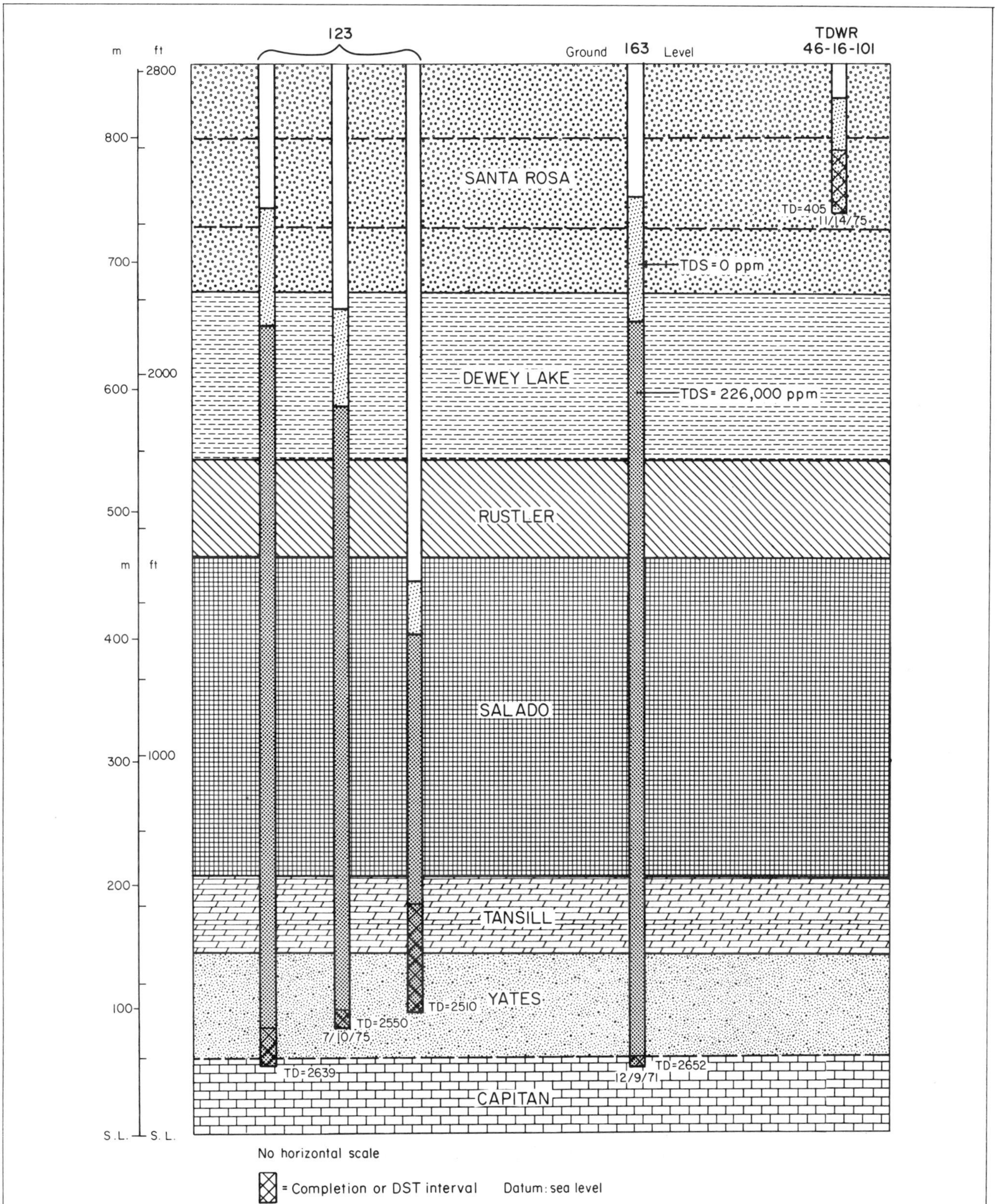

Figure 12. Schematic diagram of static fluid levels for formations in southwest Winkler County, Texas. All tests in well 123 were made in same borehole at different depths. Theoretically, pure water (TDS = 0 ppm) should rise higher than Salado Formation; water saturated with respect to sodium chloride (TDS = 226,000 ppm) should rise at least to Salado Formation. Actual salinities are between these two extremes. Depths of test intervals shown in table 3. Because hydraulic head in Santa Rosa Formation is higher than in other formations tested, water would flow from Santa Rosa to other formations if a permeable pathway were available. For location of wells, see figure 14.

Table 3. Formation pressure data from drill-stem tests for two wells near Winkler County sinkhole.*

Well no.	Date	Ground elev. (ft)	Drill-stem-test data: Depth of test (ft)	Formation tested	ISIP (psi)	FSIP (psi)	Height of water column (ft)†: TDS 0 (ppm)	TDS 226,000 (ppm)	Elevation of water column (ft above MSL): TDS 0 (ppm)	TDS 226,000 (ppm)
123	7/10/75	2,824	2,218-2,510	Tansill-Yates	431	402	995	854	1,455	1,314
			2,500-2,550	Yates	770	813	1,877	1,611	2,176	1,910
			2,545-2,639	Yates-Capitan	958	958	2,212	1,899	2,244	2,131
163	12/9/71	2,823	2,625-2,652	Capitan	987	987	2,279	1,956	2,464	2,141

ISIP initial shut-in pressure
FSIP final shut-in pressure
TDS total dissolved solids
MSL mean sea level

*Formation pressures are great enough to support pure (TDS = 0) or saturated (TDS = 226,000) water as high as Salado Formation (650-1,500 ft [200-460 m]). See figure 12.
†Calculated from highest shut-in pressure.

discharging at that location from the Rustler Formation increased the load of total dissolved solids in the Pecos River by at least 340 tons per day (U.S. Geological Survey, 1941). Recharge in the outcrop zone near Clayton Basin (fig. 13) maintained a hydraulic head that forced the brine (with a concentration of 125,000 to 155,000 ppm) upward to discharge points along the river. Presumably, the salt was derived from dissolution of the Salado Formation underlying the Rustler Formation.

Farther downstream, between Orla and Grandfalls, Texas (fig. 13), the salt content of the Pecos River water is increased by evaporation, transpiration of water by phreatophytes along the river channel, saline ground-water discharge, and contamination from oil wells (Pecos River Commission, 1955; Grozier and others, 1966). Samples obtained May 10 through 12, 1965, exhibited chloride concentrations of 7,710 ppm at Orla and 16,300 ppm at Grandfalls. This large increase was probably due, at least in part, to brine pollution from oil fields upstream from Grandfalls (Grozier and others, 1966). The Patton oil field straddles the Pecos River just upstream from Grandfalls. Source of salt from natural seeps was not determined.

Downstream from Grandfalls the quality of the river water gradually improves as fresher water from Cretaceous aquifers dilutes the river water. Near Imperial, Texas, 13 river miles (21 km) downstream from Grandfalls, chloride concentration from May 10 to 12, 1965, was 7,220 ppm (Grozier and others, 1966).

The number and size of dissolution features indicate that large volumes of salt have been removed from the Delaware Basin. Morgan (1942) estimates that the amount of salt removed from the Salado Formation in eastern Eddy County, New Mexico, amounts to 56 percent of all sediment and solute removed from the area during early and mid-Pleistocene. He also concludes that if present rates of discharge from aquifers in contact with the Salado continue unchanged during the next million years, the ground surface will be lowered 56 ft (17 m) over an area of 1,500 mi^2 (3,900 km^2). Anderson and Kirkland (1980) predict that over a period of 30,000 years, sufficient salt could be removed from the Delaware Basin in New Mexico alone to form 100 dissolution chambers each with a volume of 35 million ft^3 (990,000 m^3). That is, each chamber would be about 6 times larger than the Wink Sink.

DEVELOPMENT OF THE WINK SINK

Mechanisms of Dissolution

How water comes in contact with a salt body is not well understood. Anderson and Kirkland (1980) report that dissolution may occur in zones that are well removed from recharge areas of the dissolving waters. If this is true for the Wink Sink, it may be impossible to locate the source of the water that produced the dissolution chamber. Available data indicate that dissolution may have occurred by upward or downward movement of water, and that water movement may have been facilitated by the presence of an abandoned 52-year-old borehole.

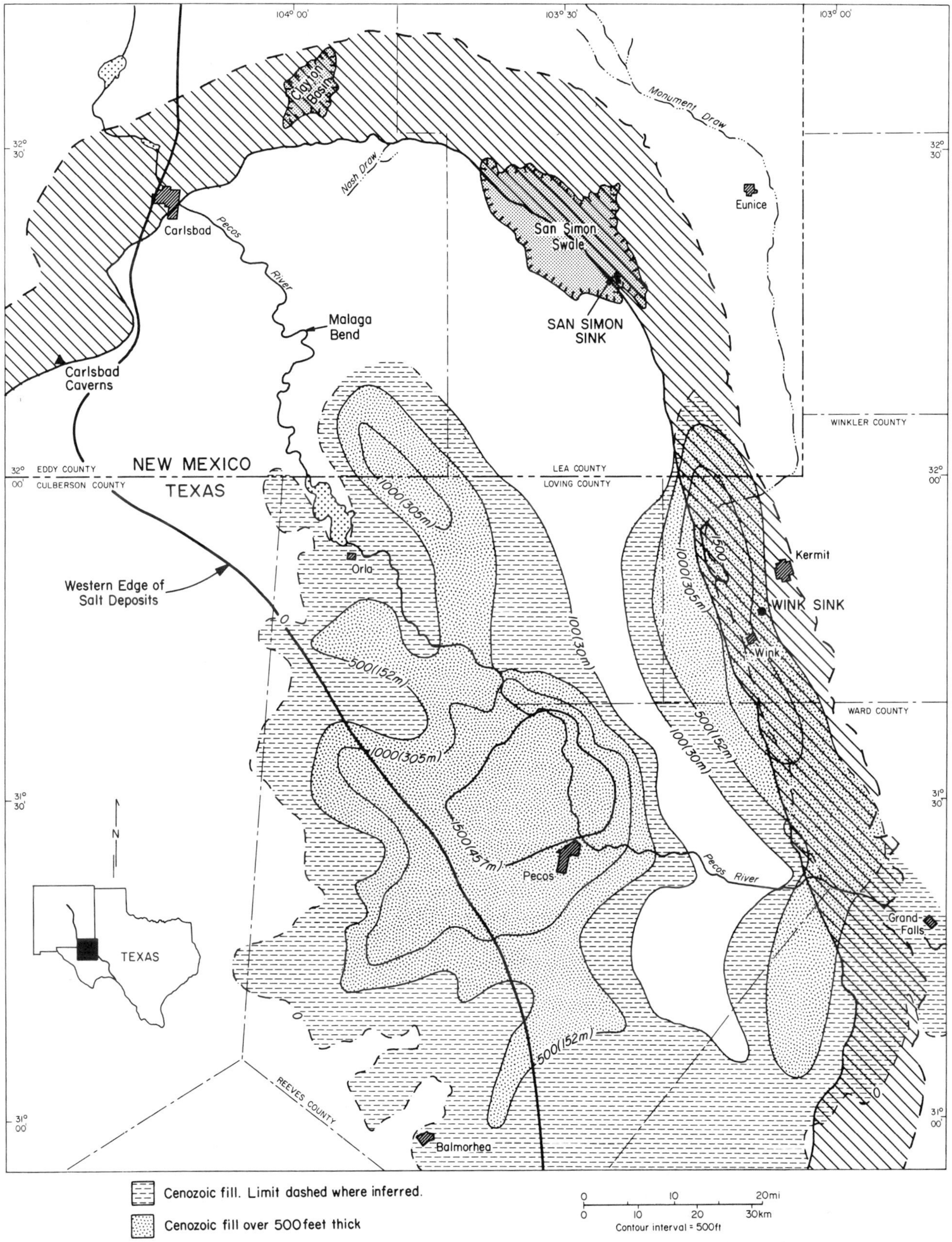

Figure 13. Dissolution and collapse features, and isopach map of Cenozoic sediments, Delaware Basin. Cenozoic sediments more than 500 ft (152 m) thick (stippled area) overlie salt dissolution zones in center of basin and along eastern side. Other dissolution features coincide with subsurface trend of Capitan Reef on northeast side of basin. Adapted from Maley and Huffington (1953), Nicholson and Clebsch (1961), Hiss (1975), and Bachman (1976).

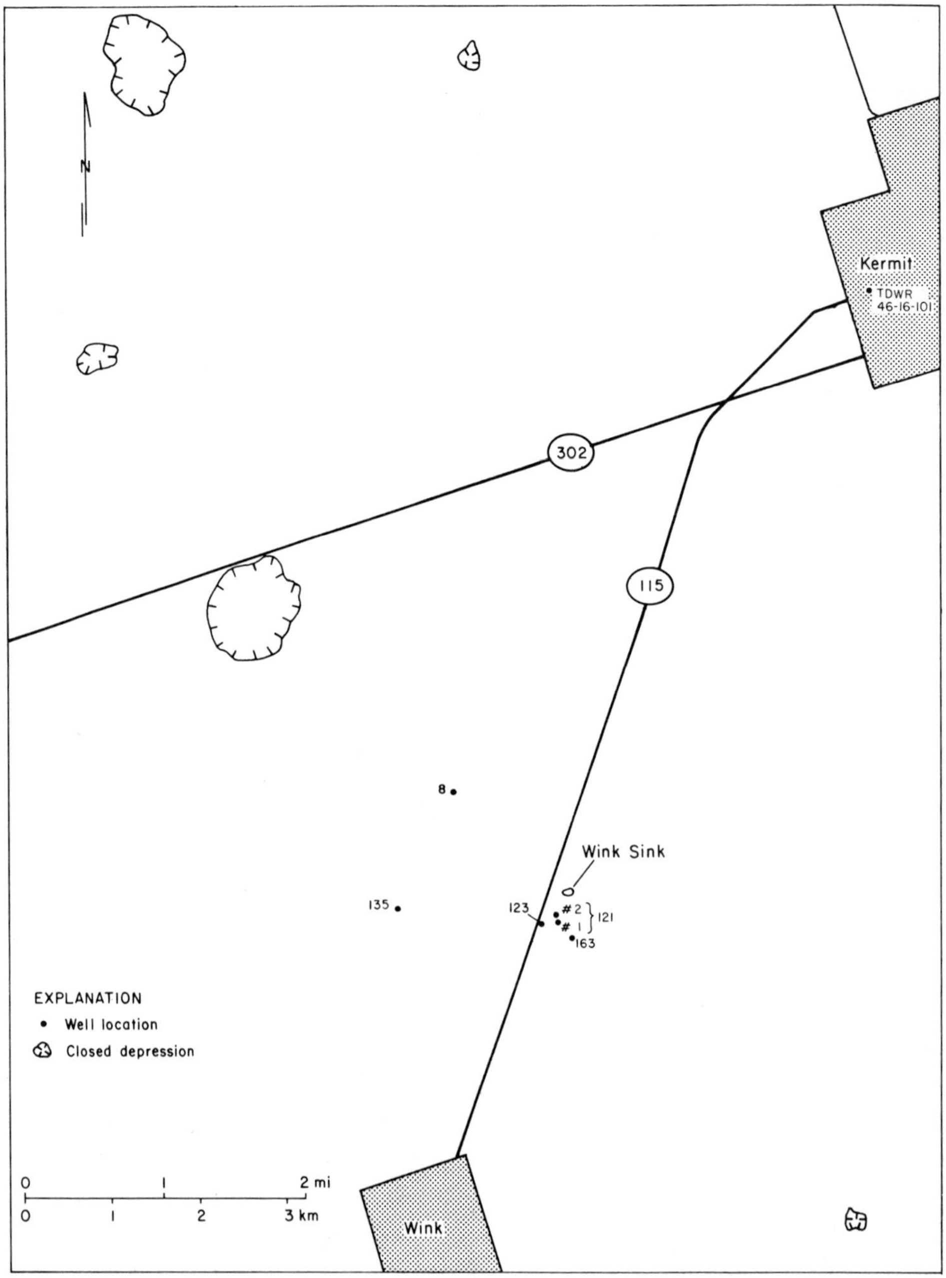

Figure 14. Evidence of salt dissolution and collapse near Wink Sink. Four wells (numbers 8; 121, hole no. 1; 121, hole no. 2; and 135) near Wink Sink lost fluids when drilled during 1927 and 1928. Loss indicates fractured or cavernous permeable conditions in vicinity. Closed depressions are probably older subsidence features, especially two small depressions north and southeast of sink. Data from wells 123, 163, and TDWR 46-16-101 used to determine hydraulic heads of formations in area (fig. 12).

Two conditions necessary for a brine-density-flow cycle exist near the Winkler County sinkhole. First, data from drill-stem tests of two oil wells within 1,500 ft (457 m) of the sink (table 3) indicate that hydraulic heads in the Yates and Tansill Formations and the Capitan Reef are at least as high as the Salado Formation (fig. 12). Historically, the head in the Capitan was higher than its present level, but withdrawal of water for waterflood projects has lowered the head in the aquifer (Hiss, 1971).

Water in the Yates Formation and in the Capitan Reef is relatively fresh. Chloride ion concentrations have been reported for water samples taken near the sinkhole (Hiss, 1975). They range from 4,300 to 35,000 mg/L for water from the Yates and from 1,300 to 3,600 mg/L for water from the Capitan. None of these concentrations approaches salt saturation levels (311,300 mg/L, or 226,000 ppm [Walters, 1978]). Thus, artesian pressure and relatively fresh water provide the first component of a brine-density-flow system.

Second, presence of permeable fracture or cavernous zones is indicated by the loss of fluid during well drilling. This occurred in four different wells drilled in 1927 and 1928 between depths of 956 and 2,293 ft (between 291 and 699 m) (fig. 14). Circulation was lost in (1) sand and red beds in the Dewey Lake Formation (well 8); (2) salt in the Salado Formation (well 121, hole number 1; and well 135); and (3) dolomite in the Tansill Formation (well 121, hole number 2). These lost circulation zones are highly permeable pathways for the movement of fluids within, above, and below the Salado Formation.

According to Anderson and Kirkland (1980), brine-density flow is now active in the Delaware Basin and has produced a "dissolution wedge" along the inner-reef margin on the eastern side of the basin (fig. 11). Most of this wedge lies west of the Wink Sink, which is directly above the reef (fig. 13). On the other hand, because the hydraulic head in the Triassic Santa Rosa Formation is higher than the head in any of the deeper aquifers (fig. 12), water would flow from the Santa Rosa downward into any of the Permian aquifers if connected by a permeable zone. Potential for downward movement of relatively fresh water may be increased by the presence of abandoned oil and gas wells that can serve as vertical pathways. Development of the Wink Sink may have been facilitated by an abandoned 52-year-old oil well, Hendrick number 10-A, which was drilled at a point within the circumference of the sinkhole. The history of the well is further discussed on page 27.

Local Geologic Setting

Stratigraphy and Structural Geology

In the vicinity of the Wink Sink, maximum thickness of Salado salt is about 700 ft (213 m) less than salt thickness in the Delaware Basin reported by Johnson and Gonzales (1978) (table 2). As illustrated by figure 7, salt beds thin locally from east to west as a result of salt dissolution. Salt thickness decreases from 945 ft (288 m) 1.2 mi (1.9 km) east of the sinkhole in well 67 to 42 ft (13 m) 2.5 mi (4 km) west of the sinkhole in well 140 (table 2). The eastern flank of a north-northwest- to south-southeast-trending solution trough (fig. 6) extends about 20 mi (32 km) north and 40 mi (64 km) south of cross section B-B′ (fig. 7).

Dissolution of salt by ground water has occurred at several levels in the Salado Formation. This is readily documented by tracing anhydrite beds laterally and by noting the progressive decrease in thickness of intercalated salt beds (figs. 7 and 15). For example, three anhydrite beds in the upper 350 ft (107 m) of the Salado can be traced between wells 66 and 139 (fig. 7), a distance of about 3.5 mi (5.6 km). Salt between the three anhydrite beds thins westward from 180 ft (55 m) to 90 ft (27 m). Between wells 87 and 81, thinning is abrupt: salt thins from 105 to 70 ft (32 to 21 m) over a distance of 2,050 ft (625 m). Two salt layers between wells 87 and 81 thin eastward, opposite of westward thinning associated with the dissolution trough (fig. 6). This thinning illustrates that dissolution can be localized.

Such abrupt variations in salt thickness most likely result from salt dissolution rather than from facies changes during deposition of the salt. Corroborating evidence for postdepositional change appears in Anderson and others (1972), who reported that horizontal laminae in the Castile and Salado Formations can be correlated for 70 mi (113 km).

Near the sink, cross section D-D′ (fig. 16) is approximately parallel to the axis of the dissolution trough. The same three anhydrite beds observed on cross sections B-B′ and C-C′ (figs. 7 and 15) occur in the upper 380 ft (116 m) of the Salado Formation, and dissolution zones are associated with all three beds. Net salt decreases southward from 600 ft (183 m) in well 10 to 415 ft (126 m) in well 109, then increases to 710 ft (216 m) in well 269 (table 2). No progressive change in salt thickness is recognized parallel to the dissolution trough, but cross section D-D′ reveals that dissolution zones extend beneath the Wink Sink.

Analysis of geophysical logs near the sinkhole indicates that dissolution of Salado salts has occurred. These logs and maps derived from log data, however, provide no evidence that permits prediction of where a sinkhole may form.

Evolution from solution cavity to collapse chimney to sinkhole may occur slowly. The development probably involves (1) roof collapse, followed by (2) gradual dissolution of the soluble part of the breccia, followed by (3) roof collapse, until a cavity becomes large enough to migrate to the surface. On the other hand, rapid development of a chimney could occur by coalescence of several solution cavities at different levels in the evaporite formation. Figures 7 and 15 illustrate superposed dissolution zones in the Salado Formation beneath the Wink Sink. These may have been precursors to the sinkhole and thus may have controlled the general location where ground collapse occurred.

Gravity Survey

A gravity survey was conducted at the site of the sink during July 9 and 10, 1980, to define the collapse zone below the sinkhole (fig. 17). The purpose of the survey was to detect density differences between the zone of collapse below the sinkhole and the undisturbed strata surrounding it. A La Coste-Romberg gravity meter accurate to 0.01 mgal was used. All gravity stations were surveyed and located horizontally and vertically to within accuracy of 0.001 ft (0.3 mm). Except for stations 42 to 47, all are marked by a concrete monument in which a nail is imbedded.

A complete Bouguer anomaly was calculated for each station using a Bouguer density (ρ) of 2.0 g/cm^3. Data for three profiles are shown in figure 17; two profiles cross directly over the sinkhole (A, B), and one lies approximately 500 ft (152 m) south of the sink (C). All three profiles show a smooth gradient without significant perturbation near the sinkhole. The

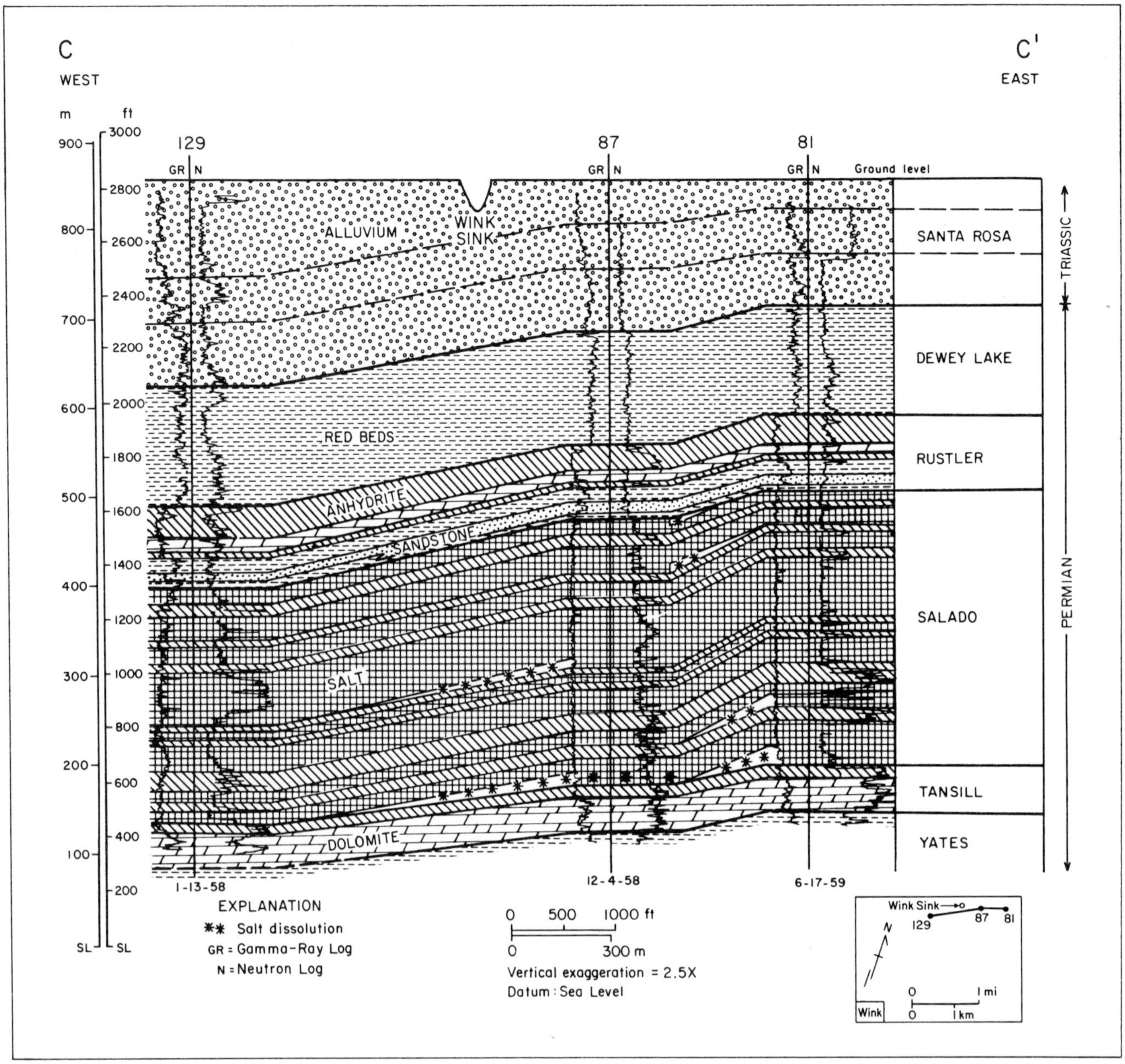

Figure 15. Detailed east-west cross section C-C′ at site of Wink Sink. Salt dissolution zones occur at several levels and can be traced laterally by noting decrease in salt thickness between adjacent anhydrite beds. Note two dissolution zones beneath Wink Sink. Number above each well refers to appendix B. Date below each well indicates date of well logging. Elevations of top and base of Santa Rosa Formation approximate. Inset shows location of line of section.

gradient is probably related to the major positive gravity anomaly associated with the Central Basin Platform east of the sinkhole (fig. 18). The two profiles oriented southwest-northeast increase eastward displaying similar slopes, whereas the profile oriented southeast-northwest decreases northward, cutting the regional gravity gradient at a slight angle. Absence of a gravity anomaly near the Wink Sink concurs with Weart's (1980) observation that collapse features in southeastern New Mexico failed to exhibit significant density contrast.

Absence of a detectable gravity anomaly related to the sinkhole is not inconsistent with expected subsurface geology. Two phenomena might yield horizontal density contrasts that could be detected with the gravity technique: (1) brecciation above a zone of collapse yielding a negative density contrast, and (2) collapse of overlying sedimentary rocks (ρ = 2.55 g/cm^3) into a void in the Salado Formation (ρ = 2.40 g/cm^3) yielding a positive density contrast. A zone of collapse in the Salado Formation exactly the size of the sinkhole would have a positive gravity anomaly of only 0.04 mgal (calculation made using vertical cylinder model developed by Dobrin [1960]). A zone of collapse and brecciation above the Salado would be expected to yield a negative anomaly of similar

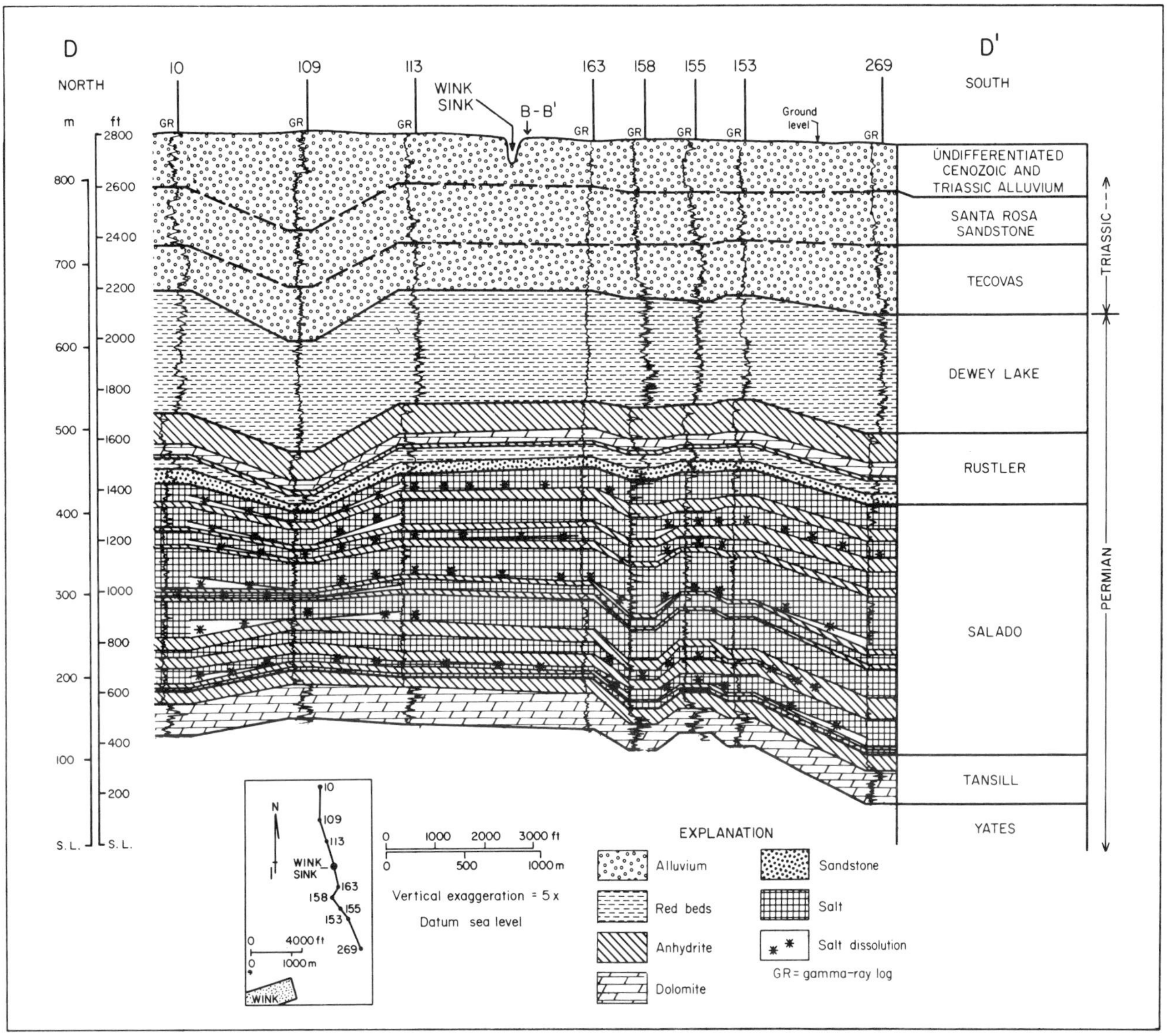

Figure 16. Local north-south cross section D-D′ at Wink Sink. Several dissolution zones are documented including two beneath sinkhole that do not appear on cross section B-B′ (fig. 7). Elevations of top and base of Santa Rosa Formation approximate. Inset shows location of line of section. Number above each well refers to appendix B.

magnitude. Thus, each phenomenon could produce a density contrast that neutralizes the other's effect, resulting in no anomaly. In addition, inaccuracies in determinations of Bouguer density, elevation, and latitude of gravity stations could approximate 0.0446 mgal (Speed, 1970), making detection of a 0.04 mgal anomaly impossible. Gravity data indicate that no significant void space remains below the sinkhole, a conclusion corroborated by first-order leveling surveys, which show that subsidence around the sinkhole is decreasing with time.

First-Order Leveling Survey

Professional surveyors conducted first-order leveling surveys at the Wink sinkhole to monitor movement of the ground surface. An AGA-brand Geodimeter was used to measure to an accuracy of 0.001 ft (0.3 mm); nails set in concrete served as survey monuments. Each survey documented changes in elevation relative to a concrete monument outside the area affected by sinkhole development (fig. 19A to D). Horizontal distances between monuments were established in relation to two stable points 450 and 600 ft (137 and 183 m) beyond the boundary of the subsided area. Latitude and longitude of the two horizontal control points were taken from the Wink North quadrangle map (U.S. Geological Survey, 1970), and the bearings of the survey monuments established by the surveyors are based on this control.

Results from surveys on July 19, August 24, October 7, and December 12, 1980, show that the south side of the sinkhole settled more than other areas bordering the sinkhole (fig. 19). Maximum total subsidence

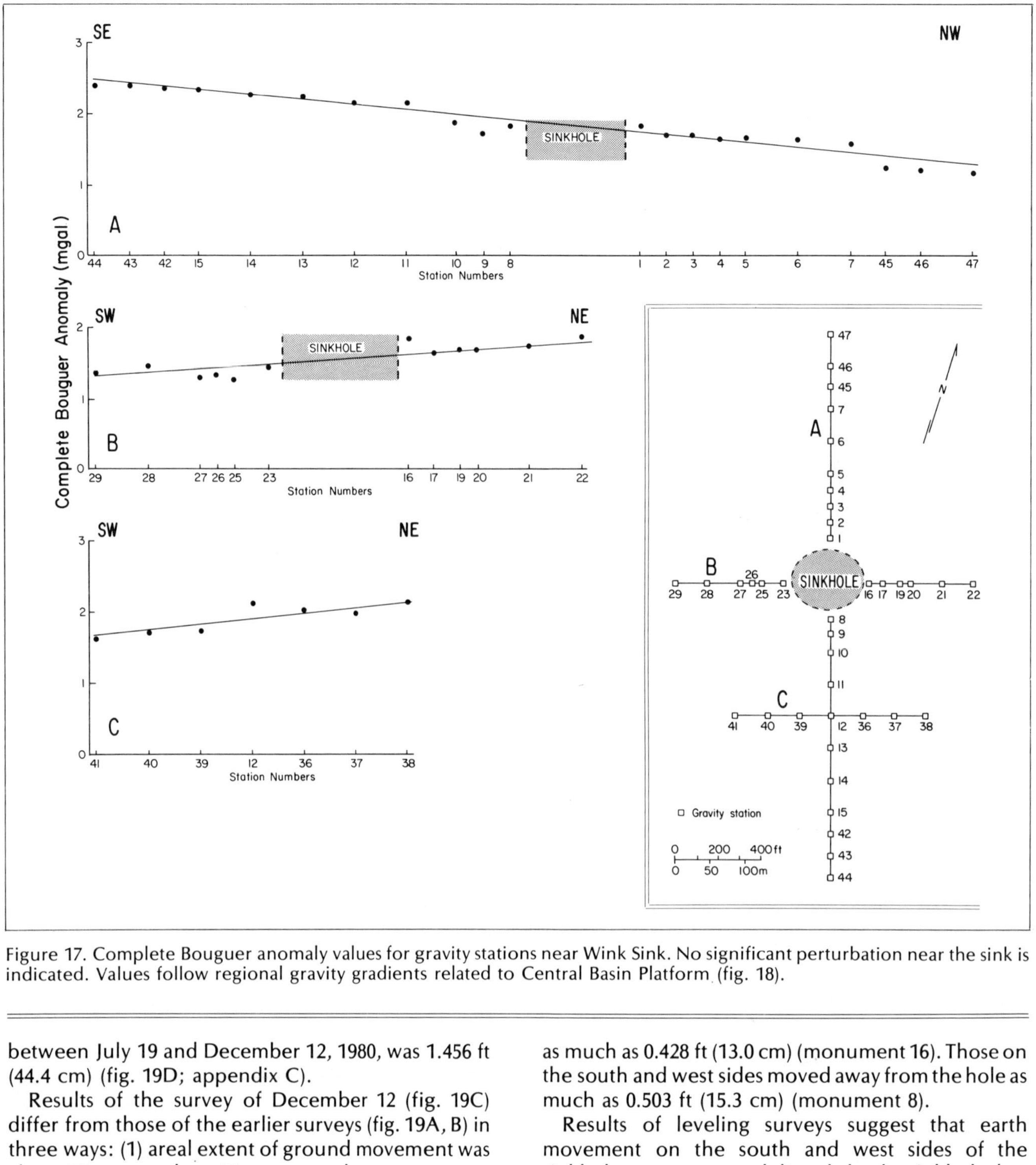

Figure 17. Complete Bouguer anomaly values for gravity stations near Wink Sink. No significant perturbation near the sink is indicated. Values follow regional gravity gradients related to Central Basin Platform (fig. 18).

between July 19 and December 12, 1980, was 1.456 ft (44.4 cm) (fig. 19D; appendix C).

Results of the survey of December 12 (fig. 19C) differ from those of the earlier surveys (fig. 19A, B) in three ways: (1) areal extent of ground movement was about 75 percent less; (2) no upward movement was observed; and (3) all subsidence was nearly concentric to the sinkhole. Ground surface more than 200 ft (61 m) away from the edge of the sinkhole was stable between October and December, 1980.

Between July and December, 1980, the horizontal distance between monuments changed only for the six nearest the sinkhole: 1, 2, 8, 9, 16, 23 (for location of monuments, see fig. 19A). Monuments on the north and east sides of the sinkhole moved toward the hole as much as 0.428 ft (13.0 cm) (monument 16). Those on the south and west sides moved away from the hole as much as 0.503 ft (15.3 cm) (monument 8).

Results of leveling surveys suggest that earth movement on the south and west sides of the sinkhole was not caused directly by the sinkhole, but that it resulted from either subsidence into minor residual void space below the sinking area, or compaction of the fractured, subsided area and closing of ground cracks that opened prior to the July 19, 1980, survey, or both. Movement was dominated by rotational slippage along curved surfaces inclined toward the sinkhole; pressure ridges up to 1.31 ft (40 cm) high south of the sink indicate that horizontal compression was occurring.

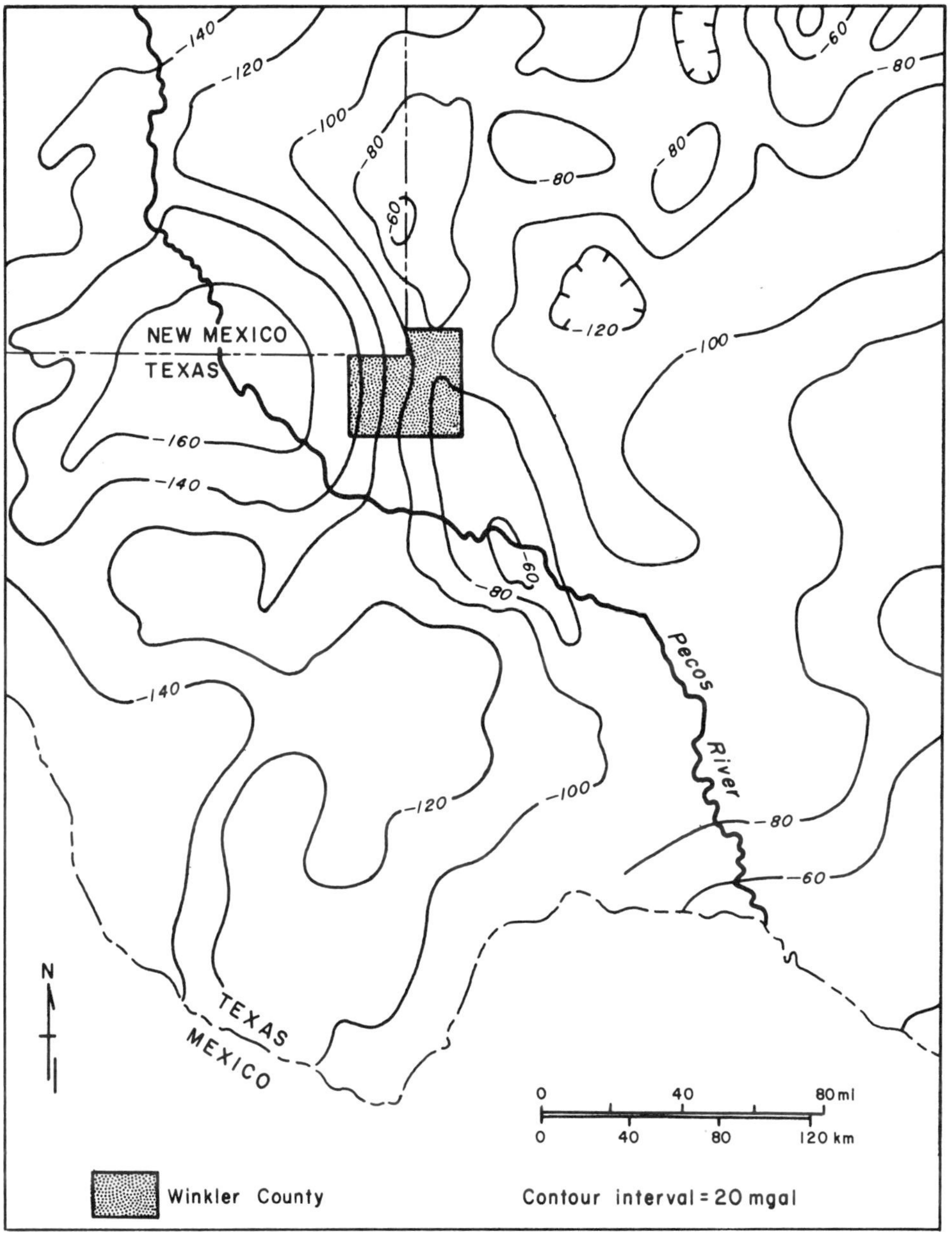

Figure 18. Bouguer anomaly values for West Texas and eastern New Mexico. Contour values are negative. Note gravity high associated with Central Basin Platform; in western Winkler County, the regional gravity anomaly increases from west to east. Adapted from Woollard and Joesting (1964).

By contrast, earth movement on the north and east sides of the sinkhole was dominated by planar movement. This type of slope failure has been described by Embleton and Thornes (1979) as slab failure. Steeply inclined tension fractures separate blocks from the surrounding undisturbed material. The blocks are thin in relation to height and tilt toward the hole. Failure occurs by toppling as blocks slide or break along a plane inclined toward the hole.

Subsidence was accompanied by cracking of the ground surface (fig. 3) as subsiding areas closer to the hole separated from peripheral areas. Cracks are most abundant south of the sinkhole, but occur on all sides of the sink and cover an area 740 ft (225 m) in diameter. Separation along the cracks measures as much as 6 inches (15 cm), although soil slumping from edges of the cracks increases the apparent width to as much as 24 inches (61 cm) (fig. 3B).

Tension fractures concentric with the sinkhole have widened more than tension fractures tangent to the sinkhole, as a photograph of the area southeast of the sink illustrates. The crack shown in figure 3A did not widen between June 27 and November 18, 1980, although one concentric crack monitored during the same period increased from a width of 3.5 inches (9 cm) to 5.6 inches (14 cm).

The leveling surveys revealed an unexpected phenomenon: upward movement outside the subsiding area surrounding the sinkhole (fig. 19A, B). Maximum upward movement of 0.223 ft (6.80 cm)

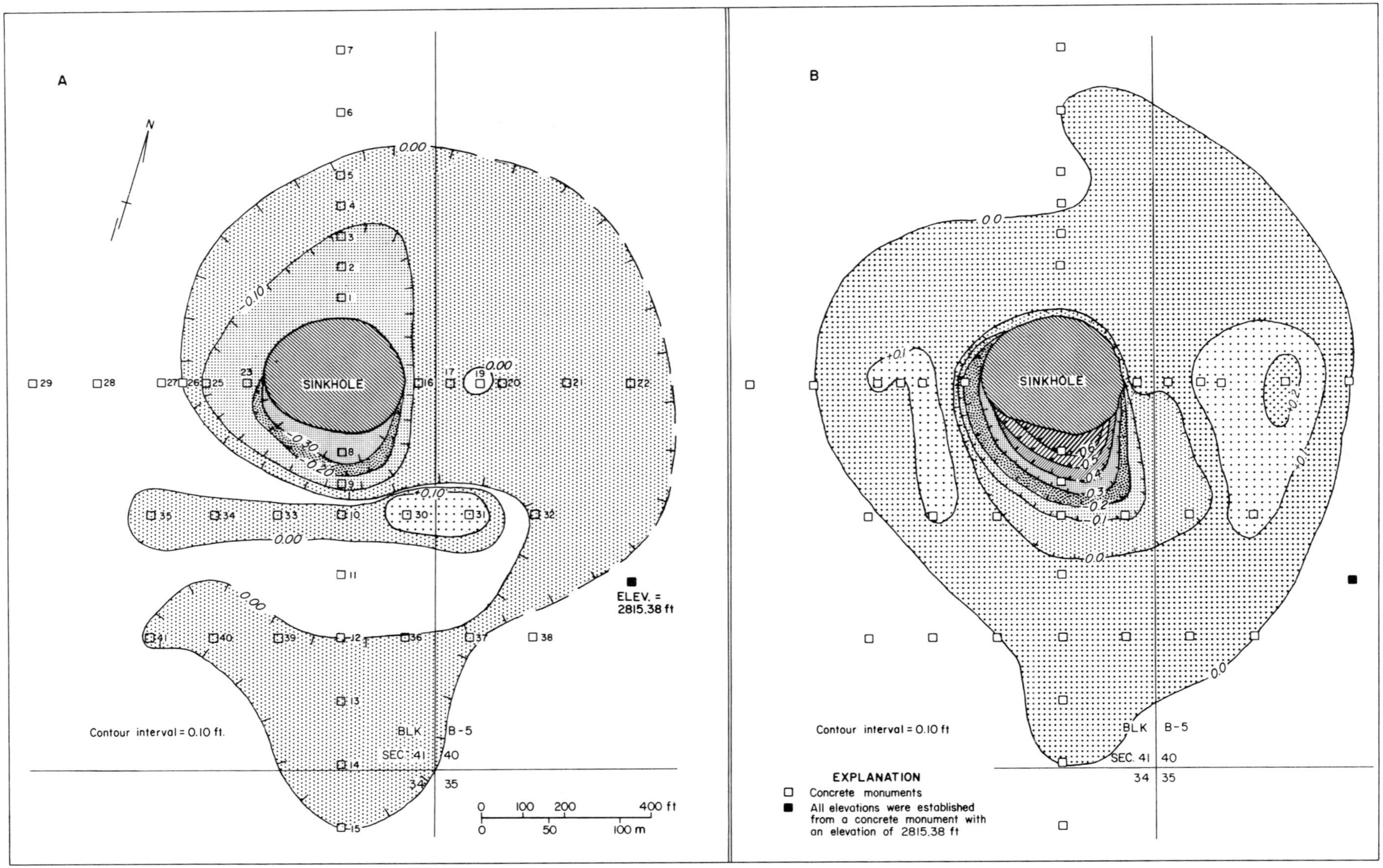

Figure 19. Changes of surface elevation at Wink Sink taken from results of first-order leveling surveys. Maximum subsidence occurred on south side of sinkhole: (A) July 19 to August 24, 1980. Monuments 18 and 24 destroyed prior to completion of surveys. Limit of subsidence dashed where inferred. (B) August 24 to October 7, 1980.

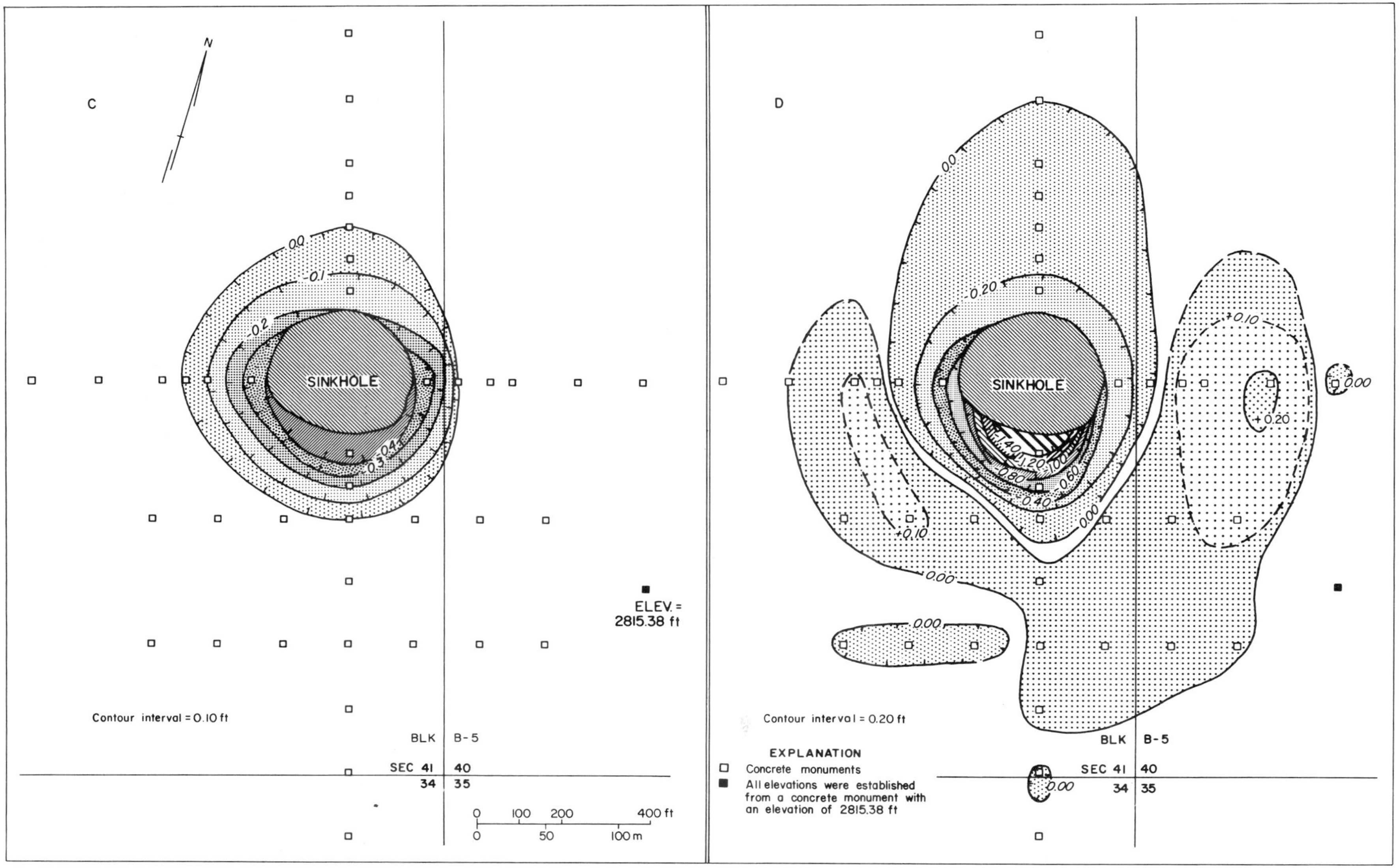

figure 19 (con.)
(C) October 7 to December 12, 1980. Note dramatic decrease in area of subsidence. (D) Summary: July 19 to December 12, 1980. Total maximum subsidence exceeded 1.4 ft (42.7 cm) on south side of sinkhole. Two areas flanking sinkhole rose slightly during this period. Limit of uplift dashed where inferred. For monument numbers, see figure 19A.

Table 4. Chemical analyses of water samples from Wink Sink and nearby wells.*

	Wink Sink	Water Well 1	Water Well 2	Water Well 3
Aquifer (Depth in ft)	—	Qal/SR 200	SR 248	Qal 200
Date collected	6/7/80	10/15/70	12/12/68	2/21/69
Date analyzed	6/9/80	10/27/70	1/28/69	3/12/69
Calcium	840	890	600	870
Magnesium	158	195	138	143
Sodium and/or Potassium	935	730	630	670
Sulfate	1,674	2,070	1,250	1,220
Chloride	2,024	1,660	1,440	1,940
Iron	0.68	No data	No data	No data
Silica	No data	33	22	39
Total hardness as $CaCO_3$	2,750	3,020	2,080	2,750
Bicarbonate	229	115	62	153
TDS	5,860	5,600	4,111	4,958
Hydrogen sulfide	0.0	No data	No data	No data
pH	7.46	7.6	7.2	7.1

Qal Quaternary alluvium
SR Santa Rosa Formation
TDS total dissolved solids

*All chemical analyses except pH reported in mg/L.
Source: Texas Department of Water Resources (1956-1979).

occurred at monument 21 between August 24 and October 7, 1980 (appendix C). During the same period, 17 of the 39 monuments moved upward on all sides of the sinkhole (fig. 19B). One might suspect that the benchmark from which the other elevations are established had moved downward, giving the false impression that other monuments had moved upward. However, several monuments displayed no movement relative to the benchmark, making that explanation implausible.

Hydrology of the Wink Sink

Water-Level Changes

When the sinkhole began to form on June 3, 1980, the top of the water surface was about 33 ft (10 m) below the ground surface on the northwest side of the sinkhole. By June 6, the water surface was 3 ft (0.9 m) lower. Three weeks later, on June 27, the water level had stabilized at about 66 ft (20.1 m) below the ground surface. On July 10 and November 18, the water level was still about 66 ft (20.1 m) lower than the ground surface northwest of the sinkhole, the most stable side of the sinkhole.

Ground water was elevated above the water table as the roof of the solution cavity collapsed and displaced the water. The rapid drop of the water level in the sinkhole was probably caused by lateral movement of water out of the cavity into the pore spaces in the surrounding unsaturated zone above the water table. This lateral movement of water may have been the last step in the formation of the sinkhole. While the cavity migrated upward through saturated or impermeable strata, the weight of the roof was partly supported by water in the cavity. But when the ceiling of the cavity emerged above the water table, this support was lost and the roof collapsed into the cavity, breaching the surface and displacing water upward above the local water table.

Water Chemistry

A water sample taken from the Wink Sink on June 7, 1980, was analyzed for major chemical constituents by Martin Water Laboratories, Inc., of Monahans, Texas. These results, along with results of chemical analyses of water from three nearby wells, are shown in table 4. Because water samples were not collected from wells during this study (June 3 through December, 1980), review of water chemistry is limited to analysis of samples collected by the Texas Department of Water Resources before the sink formed. Results of analyses of water from three wells north of the sinkhole (fig. 20) are shown because the water table in the alluvium tapped by the wells slopes southward near the sinkhole from a potentiometric mound west of Kermit (Couch, 1970). Thus, water in the Quaternary alluvium aquifer moves from north to south, and previous water samples taken from wells north of the sinkhole should resemble water taken from the sinkhole unless it was contaminated by waters contributed to the sinkhole from deeper aquifers.

Water from the sinkhole is higher in sodium, chloride, bicarbonate, and total dissolved solids (TDS) than the three water samples from the nearby water wells (table 4). However, the difference in TDS among these samples from or near the sinkhole is much less than the difference in TDS between these samples and those from most other nearby wells (fig. 20). The three wells immediately north of the sinkhole (table 4) are near the center of an area of high TDS values, probably resulting from ground-water contamination by oil field brines from unlined surface pits (Garza and Wesselman, 1959).

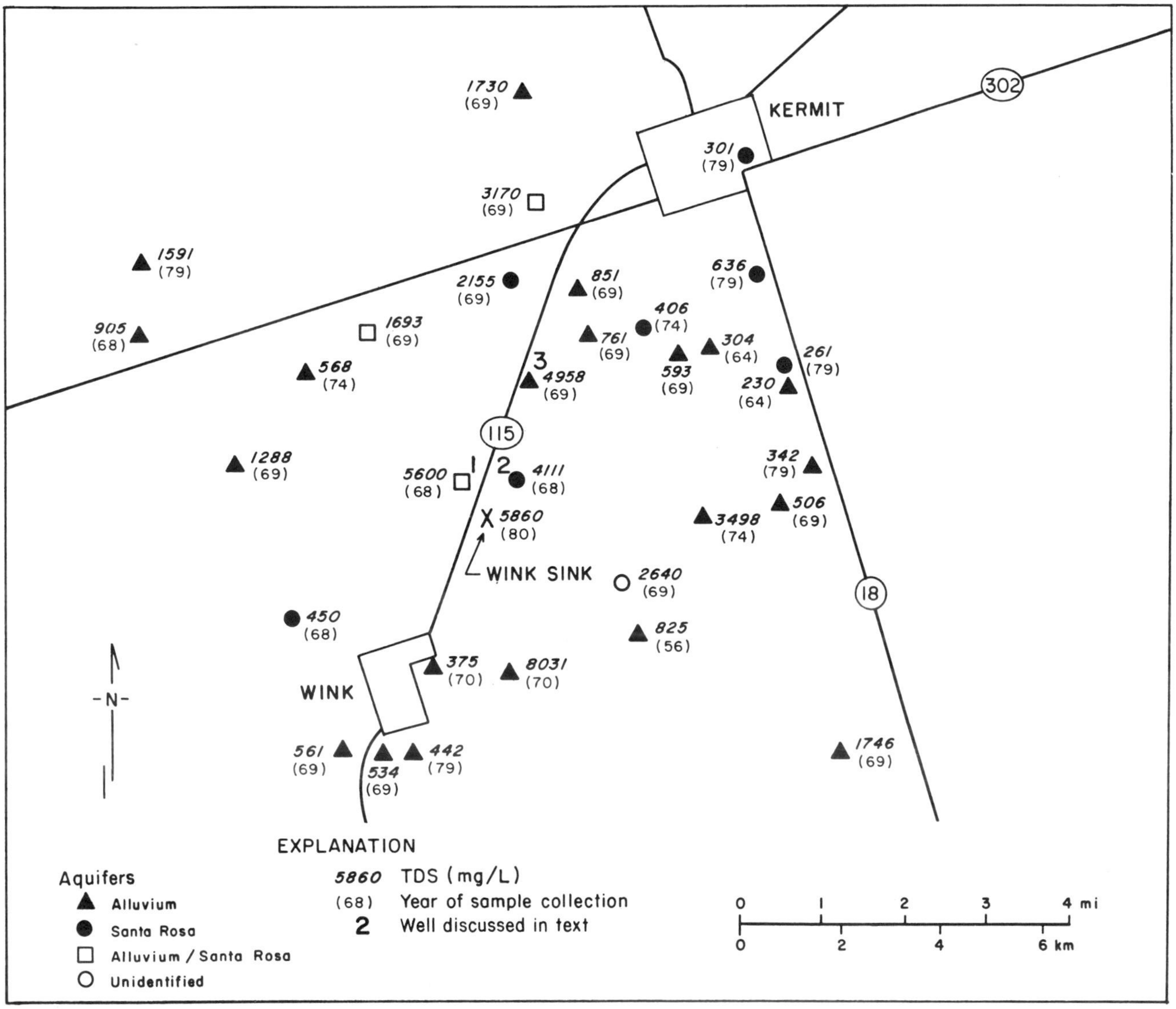

Figure 20. Total dissolved solids (TDS) in mg/L for water samples from wells in southern Winkler County. Wink Sink is near center of cluster of wells with high levels of dissolved solids. Data for water wells from Texas Department of Water Resources (1956-1979). See table 4 for summary.

Water in the sinkhole is probably a mixture of local shallow ground water and water from deeper aquifers. It is possible that some brine was transported upward from the solution cavity in the Salado Formation. To identify the source(s) of water in the sinkhole, more detailed chemical analysis, including trace element analysis, will be required. Water samples from aquifers below and above the Salado Formation, from the Salado Formation, and from oil wells producing brines should be analyzed.

HISTORY OF HENDRICK WELL NUMBER 10-A

Located within the circumference of the Winkler County sinkhole is a plugged and abandoned oil well, Hendrick well number 10-A (inset, fig. 21) (Texas Railroad Commission, 1928). The sink did not form around this borehole, but first appeared to one side of it. As the sinkhole expanded laterally by slumping and caving of the sides, the surface casing was apparently incorporated in the slump material. No eyewitnesses reported seeing the surface casing of the well as the sinkhole expanded.

Republic Production Company began drilling Hendrick well number 10-A on June 29, 1928, and completed it October 25 of the same year. The driller's logs document drilling procedures (Texas

Figure 21. Hendrick well number 10-A, section 41, block B-5, PSL survey, Winkler County. Condition of well shown as recorded when abandoned in 1964 (Texas Railroad Commission, 1928, 1930, 1964). Stratigraphic section is from area below Wink Sink (fig. 16). Note that base of Santa Rosa is close to base of surface casing, and four dissolution zones are present here in Salado Formation. For explanation of lithic symbols, see figure 16. Inset shows location of well relative to Wink Sink.

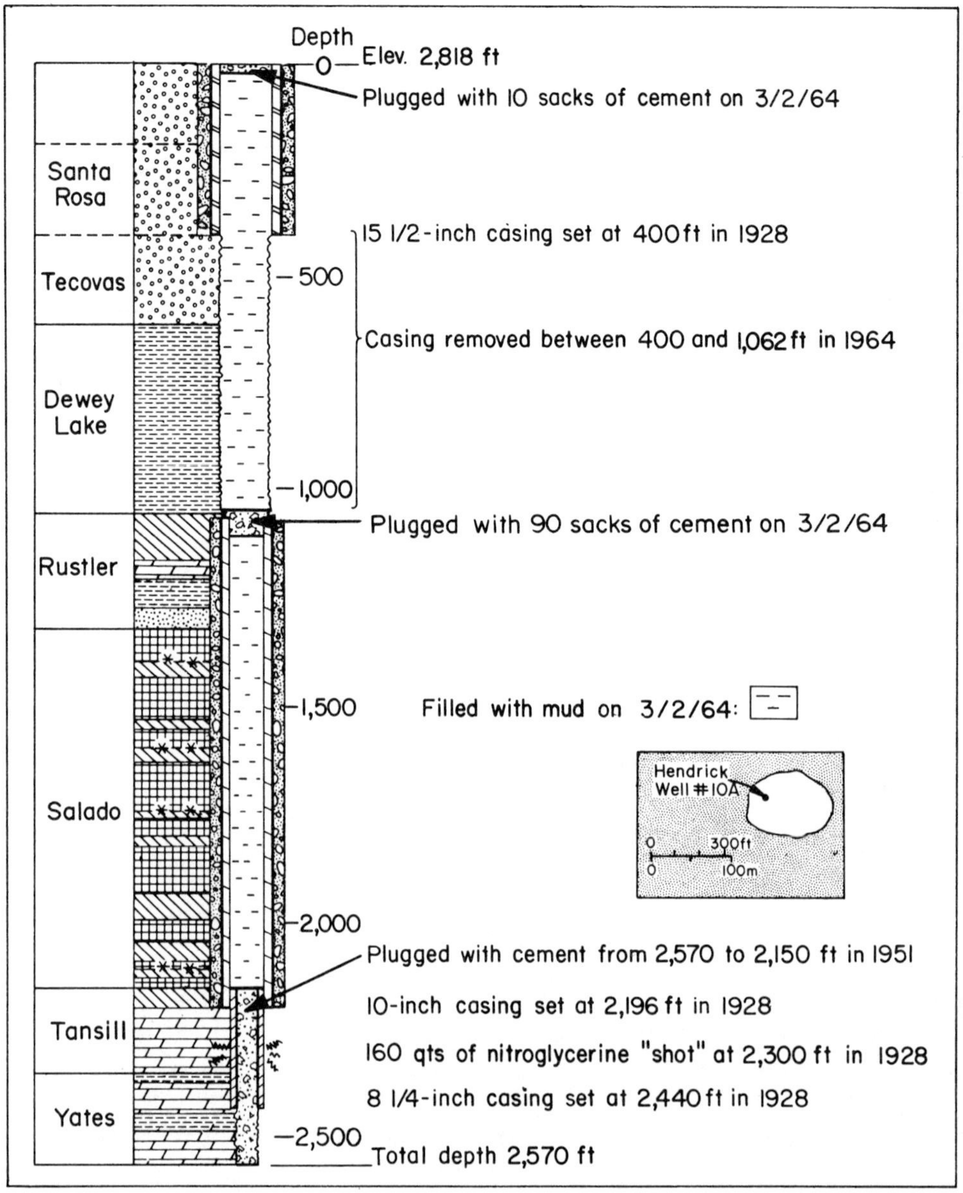

Railroad Commission, 1928) (table 5). The well was drilled with rotary tools to the top of the "brown lime of the Tansill Formation" at a depth of 2,193 ft (668 m), and cable tools were used thereafter. Surface casing 15.5 inches (39 cm) in diameter was set at a depth of 400 ft (122 m) and cemented with 300 sacks of cement (fig. 21). Ten-inch (25.4-cm) casing was set at a depth of 2,196 ft (669 m) and cemented with 800 sacks of cement. Finally, casing 8.25 inches (21 cm) in diameter was set at a depth of 2,440 ft (744 m) but was not cemented. The well was completed in the Yates Formation at a depth of 2,552 ft (778 m). No casing was set below 2,440 ft.

When the borehole deviated too much from the vertical, explosives were used to fracture the rock to allow the hole to be re-aligned. At a depth of 2,300 ft (700 m) the hole was straightened by exploding 160 qt (151 L) of nitroglycerine in the borehole, a common practice during that era of oil well drilling. Explosions could have fractured the cement lining the borehole, creating avenues for water movement.

Republic Production Company deepened the well to 2,570 ft (783 m) in January, 1930, and filed an application to deepen the well to 3,100 ft (945 m) in December, 1931 (Texas Railroad Commission, 1930, 1931). However, no drilling log on file at the Texas Railroad Commission indicates that the well was drilled deeper than 2,570 ft depth (table 5).

When the Bradberry and Sasser Company plugged the well in 1951, the well was sealed with cement from 2,570 to 2,150 ft (783 to 655 m). The wellbore was filled with mud and plugged again from 400 to 370 ft (122 to

Table 5. Driller's log for well number 10-A, T. G. Hendrick lease, section 41, block B-5, Public School Lands survey, Winkler County, Texas.*

Lithology	Color	Hard or soft	Top (ft)	Bottom (ft)	Amount (ft)
Surface sand	red	soft	0	211	211
Sand rock	brown	soft	211	250	39
Sand	red	hard	250	404	154
Red beds and broken sand	red	medium	404	474	70
Sand	red	hard	474	594	120
Red beds and sand	red	medium	594	670	76
Sand	white	medium	670	720	50
Red beds and sand	red	medium	720	979	259
Sand	red	hard	979	1,050	71
Anhydrite	white	hard	1,050	1,268	218
Anhydrite and salt	white	hard	1,268	1,678	410
Anhydrite	white	hard	1,678	2,193	515
Lime (top of lime)	white	hard	2,193	2,198	5
Set and cemented 10-inch casing, standardized					
Lime	blue	medium	2,198	2,210	12
Show of gas at 2,198 ft					
Lime	white	hard	2,210	2,220	10
Lime	blue	hard	2,220	2,235	15
Lime	white	hard	2,235	2,295	60
Increase in gas at 2,275 ft					
Lime	white	hard	2,295	2,300	5
Shot 160 qts to straighten hole					
Lime (steel line correction)	white	hard	2,312	2,365	53
Reduced hole at 2,317 ft; more gas at 2,365 ft					
Lime	white	hard	2,365	2,428	63
Lime	gray	hard	2,428	2,450	22
Shale	blue	medium	2,450	2,460	10
Lime	gray	hard	2,460	2,525	65
Increase in gas at 2,510 ft					
Lime	gray	hard	2,525	2,552	27
Top pay 2,550 ft, estimated 5,000 bbl/d, showing 80% bottom sediment and water					
- - - - - -					
Lime	gray	hard	2,552	2,568	16
Lime	gray	soft	2,568	2,570	2

*Dashed line separates 1928 data from 1930 data.

Source: Texas Railroad Commission (1928, 1930).

113 m) with 25 sacks of cement; however, this plug was later removed. Fifteen sacks of cement were used to plug the well at the surface (Texas Railroad Commission, 1951). The well was then abandoned for 13 years.

In 1964, the Mallard Petroleum Company attempted to deepen the well. Records show that the drillers were unable to reenter the hole "because of junk" in the borehole (Texas Railroad Commission, 1964). The well was plugged March 2, 1964, with 50 sacks of cement at 1,100 ft (335 m), with 40 sacks at 1,060 ft (323 m) (within the Rustler anhydrite), and with 10 sacks at the surface (fig. 21). More than 600 ft (183 m) of 10-inch (25.4-cm) diameter pipe were removed, leaving an unlined borehole (presumably filled with mud) between 1,062 and 400 ft (324 and 122 m) depth, or from below the top of the Rustler Formation to below the bottom of the Santa Rosa Formation.

No geophysical log is available for this well, but a driller's log filed by the Republic Production Company describes the strata encountered in the well (Texas Railroad Commission, 1928, 1930). This description (table 5) is very general and should be compared with the stratigraphy shown in figure 21, which is based on gamma-ray logs from wells 113 and 163 (fig. 16; appendix B).

The data from the driller's log and gamma-ray logs indicate similar depths for two distinct lithologic boundaries in Hendrick well number 10-A. The first anhydrite was encountered at a depth of 1,050 ft (320 m), and the top of a "lime" formation was recorded at a depth of 2,193 ft (668 m) (table 5). As illustrated in figure 21, depth to the first anhydrite (Rustler Formation) is 1,050 ft (320 m), and depth to the first dolomite in the Tansill Formation is 2,200 ft (670 m). Because similar depths have been recorded by both methods, we are confident that the driller correctly noted where the 10-inch (25.4-cm) casing was set and that the borehole was lined with casing through the entire Salado Formation.

The driller did not record a loss of drilling fluids, but four dissolution zones are present in the Salado below the Wink Sink (figs. 16 and 21). These zones could have formed before or after Hendrick well number 10-A was drilled as a result of ground-water movement unrelated to the presence of the borehole.

On the other hand, the abandoned well may have influenced the development of the dissolution zones and the sinkhole. Initial production from the well was estimated to be 5,000 barrels per day (bbl/d), of which 80 percent was water (Texas Railroad Commission, 1928). Pumping large amounts of saline water from this well may have increased corrosion of the pipe that lined the borehole. Water from the Yates Formation near the well has chloride ion concentrations ranging from 4,300 to 21,000 mg/L (Hiss, 1975).

Leaks are present in the casing of a nearby well of similar age. Casing in Hendrick well number 3-A, 660 ft (200 m) south of Hendrick well number 10-A, was installed in 1928. Initial production from that well was about 5,000 bbl/d, of which 90 percent was water (Texas Railroad Commission, 1928). An attempt to circulate cement behind the casing in well number 3-A failed in early June, 1980 (prior to the formation of the Wink Sink), because of leaks in the casing (Mike Handren, Petro-Lewis Co., personal communication, August 14, 1980). Presumably, these leaks were caused by corrosion. The similar production histories and ages of both wells suggest that the casing in well number 10-A may also have been perforated by corrosion.

Perforations in the casing and fractures in the cement lining the borehole may have been pathways for movement of water either up or down the borehole. Near the sinkhole, the base of the Santa Rosa Formation, a fresh-water aquifer, is at a depth of about 400 ft (122 m) (fig. 16). A poor cement job at the base of the surface casing at 400-ft- (122-m-) depth (fig. 21) could have allowed fresh water to leak down the borehole outside the casing. In addition, the absence of cement plugs or a cement lining below a depth of 2,196 ft (669 m) in Hendrick well number 10-A during the 23-year period from 1928 to 1951 may have allowed water to move upward under artesian pressure to near the base of the Salado Formation. Use of nitroglycerine to fracture the Tansill dolomite at a depth of 2,300 ft (701 m) could have increased permeability locally, thereby increasing water movement along the borehole from the Capitan, Tansill, or Yates into the base of the Salado.

Because the hydraulic head of the Santa Rosa is higher than that of the Capitan, Yates, or Tansill Formations (fig. 12), water would flow from the Santa Rosa into any of the other three formations if they were connected by a suitably permeable pathway. A borehole acting as such a pathway could contribute to salt dissolution if the casing were perforated in the salt section.

BRINE PRODUCTION AND INJECTION

The first oil well in the Hendrick Field was drilled in February, 1926, in section 42, block B-5 of the Public School Land survey (Ackers and others, 1930). Production from the field was intense and resulted in rapid depletion of the oil reservoir (Myres, 1977). From the beginning, some oil wells pumped as much as 90 percent water, and this amount increased as time passed (Texas Railroad Commission, 1928).

Beginning in 1952, oil producers in Winkler County began to inject produced water into the production horizons in waterflood projects (fig. 22). Before 1952, waterflood projects in Winkler County used fresh water obtained from Cenozoic and Santa Rosa aquifers, and the saline-produced waters were pumped into surface pits or natural drainage courses (Texas Water Commission, 1963). Waterflooding began in the Hendrick Field in 1963 and is still in operation today (Texas Railroad Commission, 1968, 1980). The brine pipeline that was ruptured by collapse on the east side of the sinkhole carried water to a pumping station south of the sinkhole (John Fogle, Gulf Production Co., personal communication, September 12, 1980). From there the brine was pumped to the Keystone Field northeast of Kermit, Texas, for use in a waterflood project.

Waterflooding should not be confused with salt-water disposal by injection. Waterflooding is a means of secondary recovery in which water is injected into

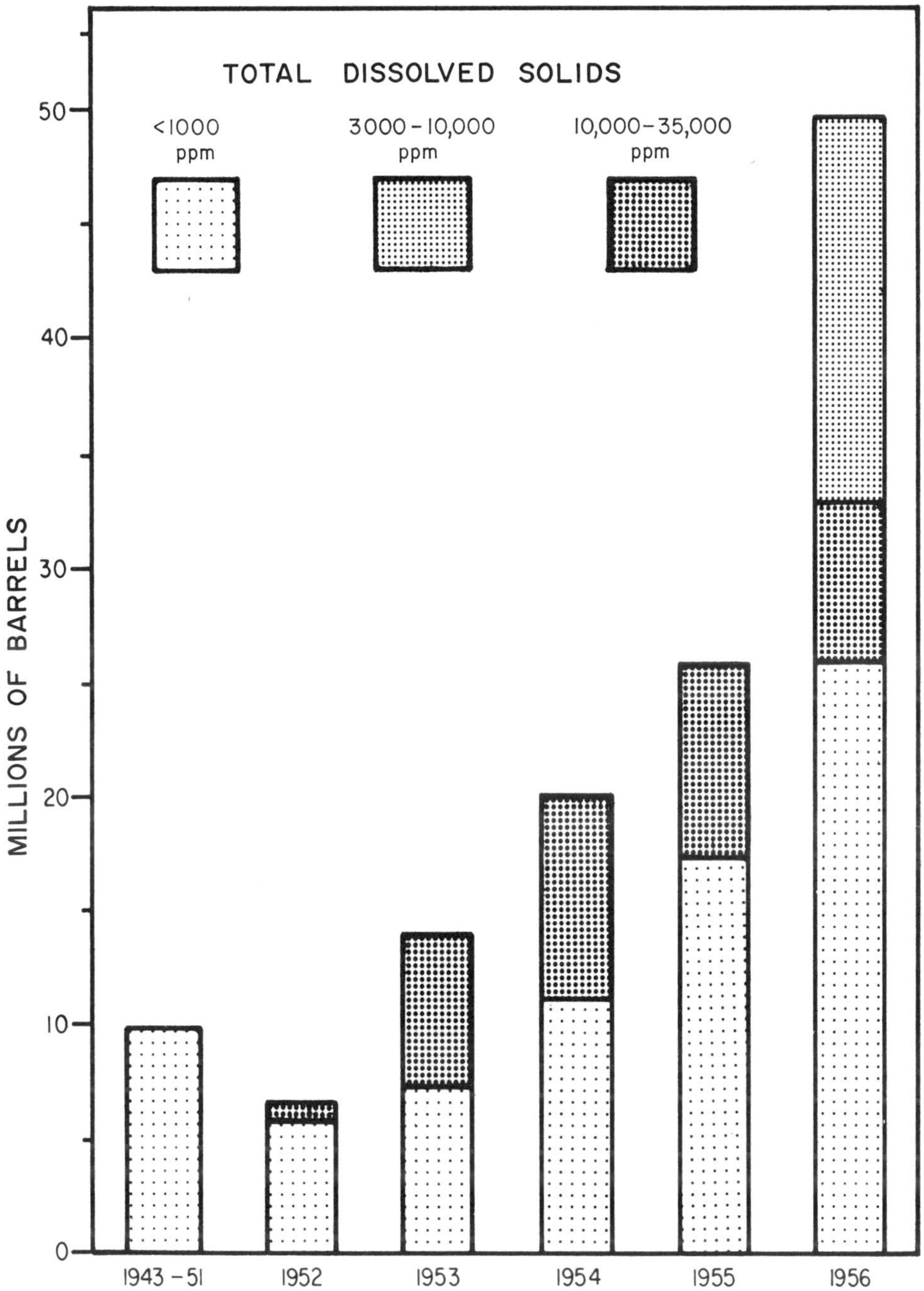

Figure 22. Water used in waterflood projects, Winkler County, Texas, 1943 to 1956. Before 1952, only fresh water (TDS less than 1,000 ppm) was used. After 1952, amount of fresh water used increased and was supplemented by more saline water. Adapted from Garza and Wesselman (1959, fig. 8).

the producing horizon to improve recovery from the hydrocarbon reservoir. Subsurface salt-water disposal is injection of water into any suitable permeable subsurface zone, normally one that already contains saline water. Disposal is designed to protect near-surface fresher water aquifers and is not intended to enhance hydrocarbon production.

In 1961, the Texas Railroad Commission conducted a statewide survey of brine production and injection (Texas Water Commission, 1963). Waterflooding and salt-water disposal were considered together as injection. According to that study, over 12 million barrels of salt water were injected in the Hendrick Field in 1961. Over half of that amount was injected into a single well about 1.7 mi (2.7 km) north of the site of the Wink Sink for a waterflood project. According to public documents, that well (T. G. Hendrick well number 22-W, block 26, section 45, PSL survey) served continuously as an injection well from 1967 to 1979 (Texas Railroad Commission, 1967 to 1979).

Applications on file with the Texas Department of Water Resources record the intervals to be used for proposed salt-water disposal wells. Until 1963, disposal into the Rustler Formation was allowed. (Since then, however, no disposal of salt water in the Rustler Formation has been approved.)

Disposal is done by gravity flow or with pressurized flow. Injection into the Yates and Seven Rivers Formations occurs by gravity flow; according to 1964 records, disposal into the Capitan Reef below a depth of 2,565 ft (782 m) required pressure of 100 psi (7.03 kg/cm^2) (Texas Department of Water Resources, 1964).

No records are kept of the amount of water disposed of into injection wells. Consequently, there is no way to determine how much water has been injected near the sinkhole. No permit was ever filed for Hendrick well number 10-A to be used as an injection well.

SUBSIDENCE FEATURES IN WESTERN WINKLER COUNTY

Aerial photographs of western Winkler County from two different years were examined to detect subsidence features that formed between 1954 and 1968. None were found. However, two closed depressions were mapped that appear to be older, degraded sinkholes on aerial photographs from 1954 and 1968 (fig. 14). Both features appear on U.S. Geological Survey topographic maps of the area as roughly circular depressions about 15 ft (4.6 m) deep. One depression is 5.6 mi (9.0 km) north of the Wink Sink; the other is 3.0 mi (4.8 km) southeast of the sink.

The diameter of both depressions, as measured on the aerial photographs, is about 930 ft (283 m), or about 2.5 times larger than the Wink Sink. However, the distance between the peripheral tension fractures surrounding the Wink Sink is about 740 ft (225 m). Larger depressions up to 3,500 ft (1,070 m) in diameter are also common features of the landscape in western Winkler County near the Wink Sink and west of Monument Draw, where numerous playa deposits previously have been mapped (fig. 4).

The two small depressions (shown on fig. 14) and the Wink Sink lie above the subsurface trend of the Permian Capitan Reef, as do a number of wet-weather ponds and depressions. Some of these internally drained depressions, such as one just south of Highway 302 (fig. 14), were formerly used for disposal of oil field brines. These features, like the Wink Sink, probably result from surface subsidence. Their orientation and location relative to the Capitan Reef suggest that the reef may have influenced their formation in the same manner that it apparently affected the dissolution preceding the appearance of the Wink Sink.

The two small depressions shown in figure 14 are probably relict sinks, and the Wink Sink may resemble their appearance in several thousand years. The depressions exhibit flat or slightly concave bottoms and gently sloping sides. Caliche zones crop out locally on the upper slopes where runoff has eroded the soil. Caliche pebbles up to 1.0 inch (2.5 cm) in diameter are found on the sloping sides of the depressions, but not on the bottoms. No vertical scarps exist at the margin of the depressions; rather, the slope of the sides of the depressions diminishes gradually over a few tens of feet until it is the same as the surrounding terrain. Bottoms of depressions have a thick grass cover, unlike the surrounding area where grass is sparse or absent.

SINKHOLES IN OTHER AREAS

Sinkholes resulting from salt dissolution have been reported in a number of places in North America including Indiana (Hall, 1976), Kansas (Walters, 1978), Michigan (Landes, 1959), Montana, North Dakota, Wyoming (Parker, 1967), South Dakota (Bowles and Braddock, 1963; Laury, 1980), Texas (Fogg and Kreitler, 1980), and Saskatchewan (De Mille and others, 1964). Sinkholes similar to the Wink Sink have been described in Saskatchewan by Gendzwill and Hajnal (1971) and in Utah by Huntoon and Richter (1979).

In Saskatchewan, a seismic reflection survey defined the shape of a collapse chimney below a circular depression 800 ft (244 m) wide known as Crater Lake (Gendzwill and Hajnal, 1971). The geophysical data showed that the chimney originated in the Prairie Evaporite (salt) Formation at a depth of 3,000 ft (915 m). Precursor to the chimney was a solution cavity 125 ft (38 m) deep and 800 ft (244 m) in diameter. The collapse chimney is about 350 ft (107 m) in diameter. When it formed, the ground surface dropped about 240 ft (73 m). The Wink Sink may have originated in the same fashion but at a shallower depth.

Huntoon and Richter (1979) described collapse chimneys in Utah that probably originated as cavities formed by salt dissolution in the Paradox and Honaker Trail Formations. The cavities propagated upward by roof collapse, and their continued growth was maintained by dissolution of carbonate breccia from overlying formations. These combined processes produced chimneys that extend upward 2,000 ft (610 m) into formations above the salt. Displacement of distinctive rock fragments from the original stratigraphic position indicates that minimum downward movement within the chimneys was about 100 ft (30 m). This compares closely to the 110-ft (33-m) depth of the Wink Sink.

CONCLUSIONS

Evaporites have been dissolving in the Delaware Basin for millions of years. Formation of the Wink Sink in June, 1980, is the most recent example of surface collapse and subsidence caused by salt dissolution. Coincidence of several surface subsidence features with the trend of the Permian Capitan Reef suggests that the reef has facilitated dissolution of the adjacent and overlying salt.

The Capitan Reef may have affected dissolution in two ways. First, differential compaction of sediments overlying the reef or faults parallel to the reef may have fractured the evaporite section, providing avenues for downward ground-water movement. Second, water under artesian pressure in the reef may have moved upward into salt beds.

Although the Wink Sink may be the result of natural processes, oil field operations in the area may be related to its formation. An abandoned oil well at the site of the sinkhole may have provided a conduit for water to come into contact with the Salado salt. Water may have moved downward from the Triassic Santa Rosa Formation or Permian Rustler Formation, or upward from the Capitan Reef into the Salado Formation. Corrosion of casing in the borehole or failure of cement plugs and lining could have facilitated vertical movement of ground water. Use of explosives to fracture rock in the Tansill Formation may also have increased permeability locally or fractured the cement lining farther up the borehole.

Between July 19 and December 12, 1980, the ground surface surrounding the Wink Sink subsided as much as 1.456 ft (44.4 cm). The extent of subsidence decreased markedly between July and December as areas farther from the sinkhole became stable. Future subsidence appears likely only within about 200 ft (61 m) of the edge of the sinkhole as it appeared in November, 1980.

Effects of brine injection and waterflooding on the formation of the sinkhole have not been firmly established. Hendrick well number 10-A, located within the sinkhole, was never used as an injection well, although nearby wells were. Injecting produced waters into the formations above and below the Salado may have altered hydrologic conditions and caused complex movement of ground water into the Salado evaporite section.

Size of the sinkhole and the depth to the salt beds in the Salado Formation are similar to other sinkholes in North America. Dissolution, brecciation, and subsidence are common characteristics of these features.

ACKNOWLEDGMENTS

The U.S. Department of Energy funded this study as part of contract number DE-AC97-80ET46615. Atlantic Richfield Company graciously granted access to their property where the sinkhole is located. Nolen Faught of Atlantic Richfield was especially helpful. Hal Dean piloted an aircraft used for overflights and aerial photography. Mike Handren of Petro-Lewis Company provided results of chemical analysis of water in the sinkhole. Jim Williamson of Winkler County Civil Defense and John Weaver of *Winkler County News* made the line-and-plummet survey possible. Many citizens in the area provided firsthand accounts of the appearance and development of the sinkhole. Discussions with several persons were also very helpful: R. L. Bassett, M. E. Bentley, J. R. Ege, T. C. Gustavson, G. R. Keller, W. W. Simpkins, and R. F. Walters.

The manuscript was first typed by B. A. Dudgeon and V. C. Zeikus; Charlotte J. Frere, under the direction of Lucille C. Harrell, typeset the final report. Illustrations were drafted by John T. Ames, Micheline R. Davis, Margaret R. Day, Margaret L. Evans, Byron P. Holbert, Jamie McClelland, and David M. Ridner, under the direction of J. W. Macon, Chief Cartographer. Editing was by Rebecca P. Lasley. Judy P. Culwell designed the cover and supervised the layout and assembly, which was done by Jamie S. Haynes. L. F. Brown, Jr., T. C. Gustavson, M. A. Bauer, G. E. Fogg, and C. M. Woodruff, Jr., reviewed the manuscript. Final reader was S. C. Caran.

REFERENCES

Ackers, A. L., DeChicchis, R., and Smith, R. H., 1930, Hendrick field, Winkler County, Texas: American Association of Petroleum Geologists Bulletin, v. 14, no. 7, p. 923-944.

Adams, J. E., 1944, Upper Permian Ochoa Series of Delaware Basin, West Texas and southeastern New Mexico: American Association of Petroleum Geologists Bulletin, v. 28, no. 11, p. 1596-1625.

Anderson, R. Y., Dean, W. E., Jr., Kirkland, D. W., and Snider, H. I., 1972, Permian Castile varved evaporite sequence, West Texas and New Mexico: Geological Society of America Bulletin, v. 83, no. 1, p. 59-86.

Anderson, R. Y., Kietzke, K. K., and Rhodes, D. J., 1978, Development of dissolution breccias, northern Delaware Basin, New Mexico and Texas, *in* Geology and mineral deposits of Ochoan rocks in Delaware Basin and adjacent areas: New Mexico Bureau of Mines and Mineral Resources Circular 159, p. 47-52.

Anderson, R. Y., and Kirkland, D. W., 1966, Intrabasin varve correlation: Geological Society of America Bulletin, v. 77, no. 3, p. 241-256.

_________ 1980, Dissolution of salt deposits by brine density flow: Geology, v. 8, no. 2, p. 66-69.

Arbingast, S. A., Kennamer, L. G., Ryan, R. H., Buchanan, J. R., Hezlep, W. L., Ellis, L. T., Jordan, T. G., Granger, C. T., and Zlatkovich, C. P., 1976, Atlas of Texas: The University of Texas at Austin, Bureau of Business Research, 179 p.

Bachman, G. O., 1976, Cenozoic deposits of southeastern New Mexico and an outline of the history of evaporite dissolution: U.S. Geological Survey, Journal of Research, v. 4, no. 2, p. 135-149.

Barnes, V. E., 1976, Pecos sheet: The University of Texas at Austin, Bureau of Economic Geology, Geologic Atlas of Texas, scale 1:250,000.

Bodenlos, A. J., 1978, Solution features in Delaware Basin evaporites (abs.), *in* Geology and mineral deposits of Ochoan rocks in Delaware Basin and adjacent areas: New Mexico Bureau of Mines and Mineral Resources Circular 159, p. 87.

Bowles, C. G., and Braddock, W. A., 1963, Solution breccias of the Minnelusa Formation in the Black Hills, South Dakota and Wyoming: U.S. Geological Survey Professional Paper 475-C, p. L91-L95.

Couch, H. E., 1970, Quality of ground water in the vicinity of Kermit, Winkler County, Texas: Texas Water Development Board, unpublished File Report 29, 38 p.

De Mille, G., Shouldice, I. R., and Nelson, H. W., 1964, Collapse structures related to evaporites of the Prairie Formation, Saskatchewan: Geological Society of America Bulletin, v. 75, no. 49, p. 307-316.

Dobrin, M. B., 1960, Introduction to geophysical prospecting: New York, McGraw-Hill, 446 p.

Embleton, C., and Thornes, J., eds., 1979, Process in geomorphology: New York, John Wiley and Sons, 436 p.

Fogg, G. E., and Kreitler, C W., 1980, Impacts of salt-brining on Palestine salt dome, *in* Kreitler, C. W., and others, Geology and geohydrology of the East Texas Basin (1979): The University of Texas at Austin, Bureau of Economic Geology Geological Circular 80-12, p. 46-54.

Garza, S., and Wesselman, J. B., 1959, Geology and groundwater resources of Winkler County, Texas: Texas Board of Water Engineers Bulletin 5916, 200 p.

Gendzwill, D. J., and Hajnal, Z., 1971, Seismic investigation of the Crater Lake collapse structure in southeastern Saskatchewan: Canadian Journal of Earth Sciences, v. 8, no. 12, p. 1514-1524.

Grozier, R. U., Albert, H. W., Blakey, J. F., and Hembree, C. H., 1966, Water-delivery and low-flow studies, Pecos River, Texas, quantity and quality, 1964 and 1965: Texas Water Development Board Report 22, 21 p.

Hall, R. D., 1976, Stratigraphy and origin of surficial deposits in sinkholes in south-central Indiana: Geology, v. 4, no. 8, p. 507-509.

Hills, J. M., 1968, Permian Basin field area, West Texas and southeastern New Mexico, *in* Saline deposits — International Conference on Saline Deposits, Houston, Texas, 1962: Geological Society of America Special Paper 88, p. 17-27.

_________ 1970, Late Paleozoic structural directions in southern Permian Basin, West Texas and New Mexico: American Association of Petroleum Geologists Bulletin, v. 54, no. 10, p. 1809-1827.

Hiss, W. L., 1971, Water-level changes in the Permian Capitan aquifer: New Mexico Professional Engineer, v. 23, no. 1, p. 16-22.

_________ 1975, Chloride-ion concentration in ground water in Permian Guadalupian rocks, southeast New Mexico and West Texas: New Mexico Bureau of Mines and Mineral Resources, Resource Map No. 4, scale 1:506,880.

Huntoon, P. W., and Richter, H. R., 1979, Breccia pipes in the vicinity of Lockhart Basin, Canyon Lands area, Utah: Ninth Field Conference, Permianland, Four Corners Geological Society Guidebook, p. 47-53.

Johnson, K. S., and Gonzales, S., 1978, Salt deposits in the United States and regional geologic characteristics important for the storage of radioactive wastes: prepared by Earth Resources Associates, Athens, Ga., for Union Carbide Corporation, Nuclear Division, prime contractor for U.S. Department of Energy, Office of Waste Isolation, Y/OWI/SUB-7414/1, 188 p.

Keller, G. R., 1980, A seismicity and seismotectonic study of the Kermit seismic zone, Texas: The University of Texas at El Paso, Kidd Seismic Observatory, Final report, prepared under U.S. Geological Survey Grant 14-08-0001-G-508, 41 p.

Keller, G. R., Hills, J. M., and Djeddi, R., 1980, A regional geological and geophysical study of the Delaware Basin, New Mexico and West Texas: 31st Field Conference, Trans-Pecos Region, New Mexico Geological Society Guidebook, p. 105-111.

Landes, K. K., 1959, The Mackinac breccia, *in* Shelden, F. D., ed., Geology of Mackinac Island and Lower and Middle Devonian south of the Straits of Mackinac: Michigan Basin Geological Society, Guidebook, p. 19-24.

Laury, R. L., 1980, Paleoenvironment of a late Quaternary mammoth-bearing sinkhole deposit, Hot Springs, South Dakota: Geological Society of America Bulletin, v. 91, no. 8, p. 465-475.

Maley, V. L., and Huffington, R. M., 1953, Cenozoic fill and evaporite solution in the Delaware Basin, Texas and New Mexico: Geological Society of America Bulletin, v. 64, no. 5, p. 539-546.

McNeal, R. P., 1965, Hydrodynamics of the Permian Basin, *in* Young, A., and Galley, J. E., eds., Fluids in subsurface environments: American Association of Petroleum Geologists Memoir 4, p. 308-326.

Mercer, J. W., and Hiss, W. L., 1978, Solution of Permian Ochoan evaporites, northern Delaware Basin, New Mexico (abs.), *in* Geology and mineral deposits of Ochoan rocks in Delaware Basin and adjacent areas: New Mexico Bureau of Mines and Mineral Resources Circular 159, p. 86.

Morgan, A. M., 1942, Solution-phenomena in the Pecos Basin in New Mexico: American Geophysical Union Transactions, v. 23, pt. 1, p. 27-35.

Myres, S. D., 1977, The Permian Basin, petroleum empire of the Southwest: El Paso, Texas, Permian Press, 624 p.

Nicholson, A., and Clebsch, A., 1961, Geology and ground-water conditions in southern Lea County, New Mexico: New Mexico Bureau of Mines and Mineral Resources Ground-Water Report 6, p. 1-123.

Parker, J. M., 1967, Salt solution and subsidence structures, Wyoming, North Dakota, and Montana: American Association of Petroleum Geologists Bulletin, v. 51, no. 10, p. 1929-1947.

Pecos River Commission, 1955, Initial development, water salvage and salinity alleviation programs, Pecos River basin: Carlsbad, New Mexico, 18 p.

Speed, R. C., 1970, Gravity anomalies from cavities in salt beds, *in* Rau, J. L., and Dellwig, L. F., eds., The third symposium on salt: Cleveland, The Northern Ohio Geological Society, v. 2, p. 367-380.

Texas Department of Water Resources, 1956-1979, Water chemistry analyses: unpublished data on file at the Surface Casing Division, Texas Department of Water Resources, Austin, Texas.

_________ 1964, Salt water disposal permits: unpublished data on file at the Surface Casing Division, Texas Department of Water Resources, Austin, Texas.

Texas Railroad Commission, 1928, Well records: unpublished data on file at the Oil and Gas Division, Texas Railroad Commission, Austin, Texas.

_________ 1930, Well records: unpublished data on file at the Oil and Gas Division, Texas Railroad Commission, Austin, Texas.

_________ 1931, Well records: unpublished data on file at the Oil and Gas Division, Texas Railroad Commission, Austin, Texas.

_________ 1951, Well records: unpublished data on file at the Oil and Gas Division, Texas Railroad Commission, Austin, Texas.

_________ 1964, Well records: unpublished data on file at the Oil and Gas Division, Texas Railroad Commission, Austin, Texas.

_________ 1967-1979, Proration schedules for Hendrick Field: unpublished data on file at the Oil and Gas Division, Texas Railroad Commission, Austin, Texas.

_________ 1968, A survey of secondary recovery and pressure maintenance operations in Texas to 1968: Texas Petroleum Research Committee Bulletin 68, 470 p.

_________ 1980, Proration schedules for Hendrick Field: unpublished data on file at the Oil and Gas Division, Texas Railroad Commission, Austin, Texas.

Texas Water Commission, 1963, A statistical analysis of data on oil field brine production and disposal in Texas for the year 1961 from an inventory conducted by the Texas Railroad Commission: Austin, Texas.

U.S. Geological Survey, 1941, Water resources of the Pecos River basin, Report B, geology and ground water: Ground-water Division, Pecos River Joint Investigation, Part III, 93 p.

_________ 1970, Wink North topographic map: 7.5-minute map series, scale 1:24,000.

Walters, R. F., 1978, Land subsidence in central Kansas related to salt dissolution: Lawrence, Kansas Geological Survey Bulletin 214, 82 p.

Weart, W. D., 1980, Technical developments in the WIPP Program: Albuquerque, New Mexico, Sandia Laboratories, unpublished report, 37 p.

West Texas Geological Society, 1966, Oil and gas fields in West Texas — Symposium, V. 1: Publication 66-52, 398 p.

White, D. E., 1971, Water resources of Ward County: Texas Water Development Board Report 125, 219 p.

Woollard, G. P., and Joesting, H. R., 1964, Bouguer gravity map of the United States: Special Committee for the Geophysical and Geological Study of the Continents, American Geophysical Union and U.S. Geological Survey, scale 1:2,500,000.

APPENDIX A:
Description of stratigraphic units

Stratigraphic unit	Lithology	Thickness (ft)	Formation top
Cenozoic Alluvium	Unconsolidated sand, gravel, silt, clay; caliche with wind-blown sand on top.	0-1,200	Surface
Triassic			
Santa Rosa Formation	sandstone	200-500	First claystone/silt-stone or porosity break
Tecovas Formation	claystone	75-125	Base of clean sand
Permian			
Dewey Lake Formation	siltstone	300-500	Break between silt-stone and claystone (top of siltstones)
Rustler Formation		250-300 (Total)	Top of first anhydrite
	anhydrite	100-150	
	dolomite	25-50	
	anhydrite	10-20	
	clay-siltstone	50-75	
	sandstone	25-50	
	clay-siltstone	0-20	
Salado Formation		400-1,300 (Total)	Top of first salt in Salado Formation
	salt	0-900	
	anhydrite with minor dolomite (carbonate)	350-450	
Tansill Formation		75-125 (Total)	Base of last salt in Salado Formation
	anhydrite	50-75	
	dolomite with clastics (dolomitic muds)	100-150	
Yates Formation	dolomite with clastics (sandstone, shales, and dolomitic muds)	not determined	First clastic mud break

APPENDIX B:
Wells cited in the text

Well no. this study	Operator	Lease and well no.	Date of log	Survey	Block	Section
8	Republic Production Co.	T. G. Hendrick #8	—	PSL	26	45
10	Gulf Oil Corp.	Grisham-Hunter #WS-5	9/10/54	PSL	26	46
59	Frank & George Frankel	Driver #1	8/16/53	PSL	B-6	19
66	Reading & Bates Oil & Gas	Cowden "B" #2	5/22/61	PSL	B-5	39
67	Shell Oil Co.	Shell et al. Cowden A-1	10/31/58	PSL	B-5	39
77	Pan American Petroleum Corp.	Hendrick-Weeks #6	9/15/59	PSL	B-5	40
81	Pan American Petroleum Corp.	Hendrick-Weeks #2	6/17/59	PSL	B-5	40
87	Finley Co.	T. G. Hendrick #3	12/04/58	PSL	B-5	40
109	Rycade Oil Corp.	Atlantic-Hendrick E-5	7/20/57	PSL	B-5	41
113	Mallard Petroleum Co.	Shell-Hendrick #1	7/13/64	PSL	B-5	41
121	Republic Production Co.	T. G. Hendrick #2	—	PSL	B-5	41
123	Monsanto Chemical Co.	T. G. Hendrick #9	6/01/75	PSL	B-5	41
124	Gulf Oil Corp.	Grisham-Hunter #WS-5	9/10/54	PSL	B-5	41
129	Monsanto Chemical Co.	Hendrick "A" #1	1/13/58	PSL	B-5	41
135	Republic Production Co.	T. G. Hendrick #1-B	—	PSL	B-5	42
139	Tyra & Tyra	Hendrick #1	5/08/69	PSL	B-5	42
140	Logue & Patterson	Ida Hendrick #1	11/18/68	PSL	B-5	43
141	Pasotex Pipeline Co.	Butane Storage #1	2/08/65	PSL	B-5	32
146	Cactus Drilling Co.	Hendrick B #1	1/17/63	PSL	B-5	33
147	Worth Exploration Co.	Hendrick "A" #1	7/25/62	PSL	B-5	34
148	Worth Exploration Co.	Hendrick "A" #2	12/02/62	PSL	B-5	34
153	Humble Oil & Refining Co. and Monsanto Chemical Co.	T. G. Hendrick Gas Unit #1	7/05/59	PSL	B-5	34
155	Humble Oil & Refining Co.	T. G. Hendrick #13	8/14/57	PSL	B-5	34
158	Gulf Oil Corp.	Grisham-Hunter Surface Fee #WS-7	7/06/60	PSL	B-5	34
163	Stoltz, Wagner, & Brown	Hendrick #1	11/15/71	PSL	B-5	34
172	Pan American Petroleum Corp.	Hendrick-Weeks #10	9/08/59	PSL	B-5	35
174	Humble Oil & Refining Co.	Fay Hunter Hogg #1	8/09/50	PSL	B-5	21
209	Pan American Petroleum Corp.	E. W. Cowden #18	10/06/62	PSL	B-5	37
215	Pan American Petroleum Corp.	Etta L. Milmo #1	4/05/61	PSL	B-6	17
269	Saxet Oil Co.	Hendrick A/C 128 #9	8/27/76	PSL	B-5	29
273	Hunt Oil Co.	University 21-10 #1	11/27/68	ULS	21	10
274	Ralph Lowe	University 1-7	10/21/60	ULS	21	7
278	Kern County Land Co.	Waddell #1	10/31/61	PSL	40	24
389	Cosden Petroleum Corp.	S. B. Wright #1	1/05/55	PSL	40	22
403	The Texas Co.	J. A. Thomas #2	3/02/59	PSL	B-5	18
555	Cactus Drilling Co.	University D #1	7/19/65	ULS	20	12
556	Holbrook-Midland	University #1-10	8/14/68	ULS	20	10
637	Hunt Oil Co.	University 21-11 #1	5/01/71	ULS	21	11
738	Union Texas Petroleum Co.	University 8-21 #1	10/29/74	ULS	21	8

PSL Public School Lands ULS University Land Survey

APPENDIX C:

Changes in surface elevation at Wink Sink, July through December, 1980*

Station no.	Total elevation, July 19	Change in elevation			Total elevation, Dec. 12	Summary net change
		July 19-Aug. 24	Aug. 24-Oct. 7	Oct. 7-Dec. 12		
1	2,824.687	−0.158	+0.024	−0.152	2,824.401	−0.286
2	2,824.451	−0.135	+0.057	−0.061	2,824.312	−0.139
3	2,823.445	−0.110	+0.060	0.000	2,823.395	−0.050
4	2,822.871	−0.011	−0.026	0.000	2,822.834	−0.037
5	2,821.566	−0.004	-0.022	0.000	2,821.540	−0.026
6	2,822.124	0.000	0.000	0.000	2,822.124	0.000
7	2,824.355	0.000	0.000	0.000	2,824.355	0.000
8	2,813.469	−0.363	−0.603	−0.490	2,812.013	-1.456
9	2,817.896	−0.122	−0.340	−0.212	2,817.222	−0.674
10	2,818.733	+0.068	−0.097	0.000	2,818.704	−0.029
11	2,817.823	+0.006	+0.043	0.000	2,817.872	+0.049
12	2,816.380	0.000	+0.069	0.000	2,816.449	+0.069
13	2,815.699	−0.006	+0.079	0.000	2,815.772	+0.073
14	2,815.077	−0.008	0.000	0.000	2,815.069	−0.008
15	2,814.615	0.000	0.000	0.000	2,814.615	0.000
16	2,820.613	−0.009	+0.012	−0.361	2,820.255	−0.358
17	2,821.661	−0.006	0.000	0.000	2,821.655	−0.006
19	2,820.116	+0.006	+0.132	0.000	2,820.254	+0.138
20	2,820.868	−0.007	+0.173	0.000	2,821.034	+0.166
21	2,820.298	−0.004	+0.223	0.000	2,820.517	+0.219
22	2,820.257	−0.007	0.000	0.000	2,820.250	−0.007
23	2,824.189	−0.013	−0.222	-0.353	2,823.601	−0.588
25	2,826.749	−0.119	+0.154	−0.105	2,826.679	−0.070
26	2,827.512	0.000	+0.092	0.000	2,827.604	+0.092
27	2,827.320	0.000	+0.103	0.000	2,827.423	+0.103
28	2,826.644	0.000	0.000	0.000	2,826.644	0.000
29	2,827.643	0.000	0.000	0.000	2,827.643	0.000
30	2,816.863	+0.113	−0.102	0.000	2,816.874	+0.011
31	2,816.570	+0.138	−0.063	0.000	2,816.645	+0.075
32	2,816.096	−0.006	+0.165	0.000	2,816.255	+0.159
33	2,822.336	+0.068	−0.022	0.000	2,822.382	+0.046
34	2,823.981	+0.008	+0.115	0.000	2,824.104	+0.123
35	2,824.754	+0.086	−0.015	0.000	2,824.825	+0.071
36	2,815.703	−0.009	+0.056	0.000	2,815.750	+0.047
37	2,815.185	−0.005	+0.040	0.000	2,815.220	+0.035
38	2,815.940	+0.010	0.000	0.000	2,815.950	+0.010
39	2,817.063	−0.004	0.000	0.000	2,817.059	-0.004
40	2,818.781	−0.008	0.000	0.000	2,818.773	−0.008
41	2,820.560	−0.008	0.000	0.000	2,820.552	−0.008

*All data are from first-order leveling surveys. Elevations expressed in feet.